T0275542

Digital Office Complex

With ever-increasing competitive pressures, the need to reduce the time-to-value (or time-to-bail) of a "big idea" – a new product, an organizational transformation, a healthcare initiative, or a humanitarian development project – has never been greater. Unfortunately, the current digital infrastructure for vision delivery teams is woefully inadequate. Improvement opportunities lie in replacing it with one that is designed to support teams working together in a unifying manner. The focus of this book is describing the Digital Office Complex and the accompanying digital teaming and governing capabilities needed to reengineer vision delivery.

Digital Office Complex: Reengineering Vision Delivery by Transforming Teaming offers an in-depth understanding of the elements of "digital teaming and governing" and how they can be applied to vision delivery to accelerate and improve performance. The book identifies and describes the requirements for an integrated infrastructure to support: team goal management, teamwork coordination, team decision support, team "work product" support, and approval workflow capabilities in addition to "team-to-team" navigation, "team-to-team" coordination, and "team-to-team" data exchange. The aim for this book is to describe and illustrate "digital teaming and governing" practices using the Digital Office Complex to improve performance. The book goes on to a team-centric delivery method for initiatives and artificial intelligence capabilities to augment teamwork. The book concludes with critical success factors for implementation and an approach for reengineering vision delivery.

Written for people who desire to implement the next level of high-performance teaming to improve organizational performance, this book is an ideal read for management consultants, executives, strategy managers, project managers, HR managers, team leaders, team members, and students in business and engineering programs.

Digital Office Complex
Reengineering Vision Delivery
by Transforming Teaming

Brad M. Jackson and Gary L. Richardson

CRC Press
Taylor & Francis Group
Boca Raton London New York

CRC Press is an imprint of the
Taylor & Francis Group, an **informa** business

First edition published 2024
by CRC Press
2385 NW Executive Center Drive, Suite 320, Boca Raton FL 33431

and by CRC Press
4 Park Square, Milton Park, Abingdon, Oxon, OX14 4RN

CRC Press is an imprint of Taylor & Francis Group, LLC

© 2024 Brad M. Jackson and Gary L. Richardson

ISBN: 978-1-032-54491-5 (hbk)
ISBN: 978-1-032-54677-3 (pbk)
ISBN: 978-1-003-42612-7 (ebk)

DOI: 10.1201/9781003426127

Typeset in Times
by Apex CoVantage, LLC

For Lex, Hunter, Courtney, and Nic

Contents

About the Authors

Brad M. Jackson is Co-founder and CEO of cordin8, llc. The company's enterprise teaming platform, **cordin8**, is based on his vision of an "organizational operating system" and "digital group memory" that he described in a 1996 paper, "The Dynamic, Re-configurable Organization of 2025: Mastering the Interplay Between Information Technology and Organizational Design" and "cordin8: The Organizational Operating System". He has spent over 35 years researching, designing, and developing collaborative technology solutions and creating "digital teaming and governing" practices to improve organizational performance. He has worked with global clients in oil and gas, insurance, consulting, and telecommunications.

He began his career at Texaco, where he led a team to explore the use of group decision support systems and other collaborative technologies to improve teaming as part of the Total Quality Management initiative at Texaco. As assistant to the CIO, he coordinated the global executive team to set strategies and develop standards for IT.

Brad received a BS in computer science from the University of Arkansas and an MS in computer science from the University of Houston. He was recognized in 2022 as a distinguished alumni of the college of engineering at the University of Arkansas and serves on the board of the University of Arkansas Academy of Computer Science and Computer Engineering. He lives in Houston with his incredible wife, Alexis, and can be reached at brad.jackson@cordin8.com.

Gary L. Richardson retired from the University of Houston college of technology graduate project management program as the PMI Houston endowed professor. During this 16-year period, he taught PMP and project-related programs for the university plus external US and international organizations. He has taught various project management programs to audiences in Finland, Russia, Iraq, China, and South America. He previously held professional certifications as a professional engineer (PE), project management professional (PMP), and certification in earned value. In this timeframe, he produced six professional texts on the topic of project management.

During the early phase of his career, Gary served as an officer in the US Air Force, leaving as a captain. Following the military period, he held positions as a manufacturing engineer at Texas Instruments, consultant to the comptroller at the Defense Communications Agency, Department of Labor, and the US Air Force (Pentagon) in Washington, DC. Interspersed through these positions, he was a tenured professor at Texas A&M and the University of South Florida, including adjunct stints at the University of Houston and Sam Houston State University. Following this, he moved to Texaco in various IT-related management roles, then finished his industry career at a Texaco/Aramco joint venture and Service Corporation International, where he held CIO-level management positions. In 1991, he was a finalist for outstanding IT executive for the South Texas Region. During the early academic period, Gary published four technical IT-related texts and numerous technical articles.

Through this broad array of experiences, Gary was involved with over 100 formal projects of various types, which collectively provided a real-world laboratory of experience matched with an equally broad organizational function view. Throughout his career, he has observed management and technical issues frequently encountered and along with this has been an active participant in the evolution of project management techniques that have occurred over this time.

Gary earned his BS in mechanical engineering from Louisiana Tech, an AFIT postgraduate program in meteorology at the University of Texas, an MS in engineering management from the University of Alaska, and a PhD in business administration from the University of North Texas. He lives in Houston, Texas.

Preface

This book is for those interested in implementing the next level of high-performance teaming, which we call "digital teaming and governing", to improve organizational performance. With ever-increasing competitive pressures, the need to reduce the time-to-value (or time-to-bail) of a "big idea" – a new product, an organizational transformation, a healthcare initiative, or a humanitarian development project – has never been greater. The ability to impact performance depends upon the quality of knowledge being generated by teams in the vision delivery stream and the speed at which it is shared to support decision-making. Unfortunately, the current digital infrastructure for vision delivery teams is woefully inadequate.

To date, information technology has largely played a supporting role in providing email capabilities to share ideas and feedback among team members and for creating and publishing final work products using personal productivity software. However, this infrastructure impedes better teaming. Email messages and data trapped in documents scattered across the organization work against smooth coordination and collaboration within and between teams as well as making it significantly more difficult to leverage artificial intelligence.

When teams form, they must identify and agree upon objectives. Next, they must create and efficiently execute a plan to achieve them. This requires coordination and collaboration not just among team members, many of whom may be remote, but also between teams. What is needed is an integrated infrastructure that provides team goal management, teamwork coordination, team decision support, team "work product" support, and approval workflow capabilities in addition to "team-to-team" navigation, "team-to-team" coordination, and "team-to-team" data exchange.

This book is about an architected collaborative infrastructure, the "Digital Office Complex", designed to provide this support so teams can work together in a unifying manner. It consists of "Digital Offices", which are workspaces that play a similar role to a team's physical space sometimes referred to as a "War Room". A Digital Office is where the team's digital work is visible to all team members and appropriate stakeholders. It is where team members develop their objectives and actions, update their status, and develop work products collaboratively. To support problem-solving, teams use it to brainstorm and evaluate ideas anonymously – vision elements, SWOT analysis, strategic objectives, product features, risks, recommendations, and so forth.

Because a Digital Office is part of a Digital Office Complex, teams can share plans, status, and work products.

Our aim for this book is to describe and illustrate "digital teaming and governing" practices using the Digital Office Complex to enable teams to improve:

- Clarity of vision, objectives, scope, and desired outcomes.
- Engagement of stakeholders, leadership, and team members.
- Prioritization of challenges and work
- Commitment of team members to goals and teamwork plans.
- Visibility of teamwork and teamwork performance.
- Quality of products, outcomes, results, and intellectual assets.
- Adoption of new technologies and new processes by the organization.

We outline a team-centric delivery method for initiatives and artificial intelligence capabilities to augment teamwork. The book concludes with critical success factors for implementation and an approach for reengineering vision delivery.

The journey for this "big idea" began in the mid-1980s at Texaco. Gary Richardson was Director of Technology Assessment, Planning, and Research in the corporate IT organization charged with exploring emerging technologies that primarily were aimed at making IT more efficient, such as automating software development (CASE tools), next-generation database technology, and local area networks. As part of his organization, Brad Jackson created and led a group to explore the potential of a new category of software called "groupware" to support "teaming" which, at the time, was a new management practice that came along with the Total Quality Management initiative being implemented corporatewide. It was the beginning of a lifelong collegial relationship and friendship.

Lotus Development Corporation, the publisher of the Lotus 1–2–3 spreadsheet software, approached our team to become a "beta" site for a new groupware product they were developing called "Lotus Notes". We experimented with it as a group memory platform for teams to share, discuss, and store team information, e.g., "hot topics", meeting agendas, decisions, minutes, and action items.

As part of another stream of research in this space, Brad reached out to University of Minnesota (UM) behavioral science researchers Gerry DeSanctis, Scott Poole, and Gary Dickson, who had developed a group support system called SAMM to aid teams in group decision-making processes. In a joint Texaco/UM study, teams used SAMM for team meetings, JAD sessions, vendor selection projects, and strategic planning sessions. They found that SAMM improved a team's meeting processes and outcomes. When SAMM was used, members' interest and involvement went up, quiet individuals would participate, and people whose ideas were discounted by others when presented verbally got more recognition. People left meetings with clear ideas and priorities. They expanded this research with an accompanying NSF grant to conduct a longitudinal study involving 80 teams from various business units across Texaco to gain a better understanding of team-technology needs. In an unpublished 1996 paper titled "The Dynamic, Reconfigurable Organization of 2025: Mastering the Interplay Between Information Technology and Organizational Design", Brad documented his vision:

> organization that was tightly intertwined with information technology . . . a dynamic, reconfigurable network where teams can create and insert themselves into the network. When a team inserts itself, it makes standard information about the team . . . available to the rest of the network which allows others to learn from it. There are intelligent assistants built into the network to assist teams.

The paper described the "organizational operating system" and "digital group memory", which are the predecessors of the "Digital Office Complex" and "Digital Office", respectively. In 2004, Brad began collaborating with colleague and longtime business partner, Andy Kalish, in the design and development of an enterprise teaming platform, **cordin8**, based on these concepts. Screenshots from **cordin8** are used in several chapters to illustrate "digital teaming and governing" concepts.

From the early work at Texaco, we acknowledge the significant contributions of Ed McDonald, who was Texaco's chief architect and our proponent, sponsor, and collaborator along with executive support from then-CIO Jim Metzger and team members Dave Hoffman, Laura Bess McDonald, Charlita Marrs, and Ben Lanius.

It was an incredible privilege to team with researchers Gerry DeSanctis, Scott Poole, and Gary Dickson to explore the possibilities of technology support for group decision-making. We appreciate all the teams that volunteered to participate in the studies using SAMM to explore better ways of teaming, especially Sarah Reinemeyer and her team.

We value our frequent two-hour-plus Skype calls over the last 15 years with Chris Bragg, managing partner at Bahrain-based PPM4Value, on all things involving strategy execution and technology. We enjoy the thought partnership on innovation, teaming, and technology with Bryan Seyfarth, CEO at Brightline. We appreciate the opportunity to observe and test concepts for the project management framework classes taught by Walter Viali, partner at PMOToGo, LLC. We value the executive perspective of Ken Fitzgerald, who recently joined cordin8, llc., as COO after retiring as VP corporate PMO at EMC Insurance Companies.

Brad M. Jackson
Gary L. Richardson

Big Ideas

1

INTRODUCTION

Smartphones, breakthrough solutions for major health issues, and tackling how poverty is reduced are examples of "big ideas". How does one go from a "big idea" to the realization of it? It's not a single-path process. It's a journey with many paths. It's very messy. It takes significant time. It requires teams of experts. And there's no guarantee that the idea will translate into value.

The journey starts with a highly creative process to capture a vision, or an imagined future, which expresses the "big idea" in the form of textual description, stories, pictures, drawings, diagrams, videos, and notes that teams can use to guide their implementation journey. They convey how their perspective on the future is different from the current day. A great example is Vannevar Bush's description of "Memex" in his article "As We May Think", which appeared in *The Atlantic* in 1945.[1]

> *Consider a future device . . . in which an individual stores all his books, records, and communications, and which is mechanized so that it may be consulted with exceeding speed and flexibility. It is an enlarged intimate supplement to his memory.*

From the expression of a "big idea" in the form of a vision to its realization, such as a new product, transformed organization, or results from a new policy, there are countless iterations of hypothesizing, idea development, testing, and learning. The purpose of each iteration is to create work product increments and gain insights that inform the next iteration such that each successive iteration is a little less "fuzzy", thus moving the teams closer to achieving the ultimate target. An abstract interpretation of this process is shown in Figure 1.1. "Start" is shown on the far left end with a fuzzy, opaque image that represents the beginning of visioning when very little is known, or understood, in terms of exactly how it will come to be implemented. It also represents the extraordinarily high degree of uncertainty in being able to realize it successfully. Moving along the spectrum

DOI: 10.1201/9781003426127-1

FIGURE 1.1 Vision realization spectrum from opaque start to clarity at target.

to the right, the image becomes slightly less blurry, indicating both tangible and intangible assets have been created through rounds of experimentation and integration of deliverables produced to date. Knowledge derived from each iteration informs the journey's next steps, including exiting the process altogether. Approaching the far right end of the spectrum is the clear image indicating the achievement of the hope, desire, and dream, which is the "Target".

To illustrate "big ideas" and the significant challenges of realizing them are two of the most profound ones: The "Pocket Crystal" and "Amazon".

Pocket Crystal

Marc Porat co-founded General Magic to work on his "big idea" of a future where people carried a computer in their pocket and ran their lives out of it. He called the device a "Pocket Crystal". The description of it is captured in an email he sent to Apple CEO, John Sculley, in 1990 describing his vision.[2]

A tiny computer, a phone, a very personal object. . . . It must be beautiful. It must offer the kind of personal satisfaction that a fine piece of jewelry brings. It will have a perceived value even when it's not being used. It will offer the comfort of a touchstone, the tactile satisfaction of a seashell, the enchantment of a crystal. Once you use it you won't be able to live without it.

It was more than just a small phone. It was all the things you would be able to do with it. Among some of the things you could do in addition to making calls would be to read news articles, play games, and purchase plane tickets. He captured the conceptual design for it in what he referred to as the "Red Book" which was described in the 2018 documentary about General Magic as a "complete and thorough collection of the vision for this future, scenario by scenario".[3]

With investment from Apple, General Magic began developing the "Pocket Crystal" in 1990. During that era, mobile technology was bulky and expensive. It used a small screen with large buttons. Creating the vision depicted in Porat's "Red Book" required multiple teams. There was a team to "figure out" a new hardware design that would utilize "touchscreen" technology instead of buttons. There was another team for designing and developing an operating system with a graphical user interface as well as teams to create various applications, such as email, calendar, and news. Each of these development areas was challenging in its own right, as each team was working in unchartered waters. As significant of a challenge as they faced working on their respective deliverables, coordinating the testing and integration was equally as challenging because of the unpredictability of when the required deliverables from each area would become available.

Understanding that they could not commercialize and operate all of the pieces themselves, they created the General Magic Alliance to partner with hardware and telecommunication companies, including Sony, Motorola, Philips, Apple, and AT&T. General Magic would provide the operating system that would run on their partners' hardware and operate over AT&T's private network. In 1994, after four years of development, General Magic released its first product, Sony Magic Link. While it received a tremendous amount of enthusiastic media coverage, there were only 3,000 or so units sold. And those were mostly to themselves, family, and friends.

In the documentary, many of the key players, including Porat, reflected on what went wrong. From a business perspective, there were several missteps. First, they lacked organization and management. They resisted it. The engineers did not believe they needed managers. They could just make it happen. They did not understand who their ultimate customer was. They described their target customer as "Joe Sixpack" at a time when most people didn't have a home PC or email nor would they pay $800, approximately $1,500 in 2022 dollars, for it. Business people, on the other hand, were potential major email and mobile users. There was also market confusion with Apple launching Newton, which was a personal digital assistant (PDA), in the same timeframe. And they made a bad infrastructure technology bet. They invested their development to deliver on AT&T's private network instead of the emerging "Internet", and they were not able to pivot quickly.

As a result of these mistakes, General Magic failed as a business. It closed operations in 2002. The underlying vision, however, became undeniably successful. Original team members went on to Google and Apple to create Android, the iPod, the iPhone, and the Apple Watch.

Amazon

Jeff Bezos founded Amazon in 1994, naming it after the world's largest river, which reflected the vision to make the online bookstore the largest in the world. He had come across the statistic that "World Wide Web" usage was growing by 2,300% a month. Recognizing the possibilities of selling online, he began researching the possibilities of developing an "Internet" business. His research led him to a list of 20 potential products, including software, CDs, and books, that he thought might sell well via the Internet. He saw the current process inefficiencies in selling books and chose this new paradigm as the place to start. The largest physical bookstore chains could stock only a few hundred thousand books, while a "virtual" bookstore could offer millions of titles.

In July 1995, Amazon.com began selling books online. The website was rudimentary and missing relevant information, such as the publication dates. By November, the company was selling more than 100 books a day. Amazon provided something for book buyers that was unique to an online bookstore. They created an online community, allowing their customers to create online reviews. Book buyers liked being able to research books read by others before buying themselves. By September 1996, they had sales of more than $15.7 million. Within two years, they added videos and music CDs and continued to add new product categories over the coming years as part of delivering on their vision.

Unlike General Magic, which was founded four years prior, with a large and complex development scope, Amazon came along as the Internet was at the very early stage of becoming "mainstream" and a narrow development scope. Bezos researched his target market to find the top "thing" that their prospective customers buy and that Amazon could make more efficient by providing the ability to purchase it online. The team set their strategy to create a minimum platform to deliver the basic service, and they launched it in the time span of a year. What both companies had in common were teams that embarked on a journey to create something that did not exist prior. Their task was to "figure out" how they were going to go from a "fuzzy" concept to a vision realized in the form of a product.

ISLANDS OF DIGITAL TEAMS

"Where's the link to the project team's website?" Lily, CEO of Acme, one of the world's largest companies, thought to herself while she was scrolling through her inbox. She had just finished a call with a board member who was visiting and wanted to meet with her in an hour about the status of the company's digital transformation initiative. She needed to bring herself up to speed so she could provide a high-level status and be prepared to answer questions. It had always been a challenge for her to get relevant, real-time information on any of the corporate initiatives because data was organized and stored differently by each team. A team member from the digital transformation initiative sent her a link to its website several months back. She was now looking for that email. Having finally found it, she clicked on the link and was now trying to figure out how to navigate the team's site. There was a dashboard on the home page. Unfortunately, it was a static picture showing information from the last quarter. Next, she clicked on a tab for files. While staring at a long list of file folders, she was thinking, "I'll need to look in all these folders and open multiple spreadsheet files and documents to pull together what I need".

The digital transformation initiative was comprised of eight teams organized geographically. One team served as the "initiative office" to consolidate monthly and quarterly status reporting data. Each month, teams prepared reports and emailed them to one or more intermediate teams that created a combined status for each of the respective geographies. Those reports were then emailed to the "initiative office" to produce a consolidated status report for the entire initiative. Lily found the most recent report, but it was three weeks ago. For her to create a current picture of this initiative, she would need to not only locate the data for each team but also decode its formatting. It was a daunting task that would require significantly more time than she had before her meeting.

High-performance teaming is crucial for making a "big idea" a reality. An organization's capacity to build teams that have the knowledge, skills, and experience to "figure out" the inevitably difficult challenges it will face faster than its competition is the difference between the winners and losers. When teams form, they must identify and agree upon goals. They must then create and efficiently execute a plan to achieve them which requires not only coordinating work among team members but also with other teams.

Researchers Carl Larson and Frank LaFasto studied a wide variety of teams, including the Boeing 747 airplane project team, a cardiac surgical team, a Notre Dame championship football team, a Mt. Everest expedition/ British Antarctic expedition team, and the Presidential Commission on the Space Shuttle Challenger Accident as part of their research in determining characteristics of effective team functioning. Among the attributes they identified for high-performance teaming is a "results-driven structure" which,

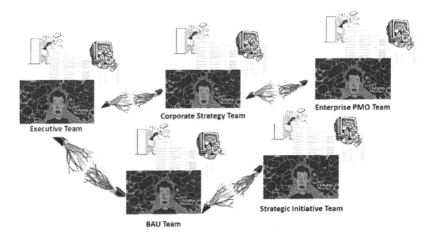

Corporate Strategy Team

Enterprise PMO Team

Executive Team

Strategic Initiative Team

BAU Team

FIGURE 1.2   Islands of Digital Teams.

in part, is a communication system where information is easily accessible and where team members accurately document issues and decisions to prevent duplication of effort and reduce confusion.[4] Information technology can facilitate this need, but its current implementation in most organizations is an impediment.

The infrastructure used by most teams today is largely based on 1980s personal productivity software, email, and file servers that automated 1950s clerical work that relied on typewriters, adding machines, inter-office memos, and filing rooms. Without reimagining an infrastructure that defines a data architecture for the organization's vision delivery process, teams are left to their own devices to make technology and data storage decisions. The result is a patchwork of technology with disparate data that the authors refer to as the "Islands of Digital Teams" and is illustrated in Figure 1.2.

Instead of a database that is accessible by all teams to enable data sharing within and between teams, each team defines and stores its data in local files that require them to extract data and email it to an intermediate team that then consolidates it into another intermediate file to email to yet another consolidation team and so on until it reaches the leadership team. Not only is it a time-consuming effort to produce this report but also, the true loss is the inability of all the teams, including leadership, to have visibility into each other's work. Not having access to the most current information makes it difficult for leaders to have a complete context of the state of execution of their strategy. It impacts the speed and quality of teamwork. It is a significant impediment to coordination that impacts project performance and strategic success.

REENGINEERING

In 1990, Michael Hammer introduced the world to business process reengineering in his seminal *Harvard Business Review* article "Reengineering Work: Don't Automate, Obliterate". Hammer wrote:

> At the heart of reengineering is the notion of discontinuous thinking – of recognizing and breaking away from the outdated rules and fundamental assumptions that underlie operations. Unless we change these rules, we are merely rearranging the deck chairs on the Titanic. We cannot achieve breakthroughs in performance by cutting fat or automating existing processes. Rather, we must challenge old assumptions and shed the old rules that made the business underperform in the first place.[5]

Remote work is among the drivers pushing to improve the underlying digital infrastructure to support teamwork. The COVID pandemic was a seismic jolt that forced a global-scale experience of "remote work", which was made possible because of collaborative technology. Before the pandemic, some companies supported employees working remotely, but most did not because of the belief that it could not be managed appropriately to be productive. However, the experience through the pandemic has proven otherwise. Organizations came to realize that they were able to be productive with people working remotely, so they have been reassessing how to conduct their business going forward. Among the reasons for doing so are the considerable cost reductions associated with relinquishing office space if employees should work from home. In many instances, there is a strong employee preference for working remotely, which can lead to a loss of talent without such a human resources (HR) policy that provides for it. However, not all employees want to work remotely. Many enjoy the social aspect of working near their colleagues. As a result, companies must decide what work arrangements will best suit them from a complete virtual model where everyone works remotely to a hybrid where employees work a few days in the office and other days remotely to everyone returns to the office.

However, independent of where a company lands on this scale, improving vision delivery performance requires reengineering the underlying technology infrastructure from the "Islands of Digital Teams" to one that is architected to enable teams to collaborate and make decisions. With OpenAI's launch of ChatGPT in November 2022, the interest in the potential of artificial intelligence (AI) applied to business areas, including portfolio and project management, has been off the charts. Roughly 80% of the time spent preparing to utilize AI is focused on gathering and cleaning data. An architected approach will better position an organization to benefit from these new AI capabilities by significantly reducing, if not eliminating, the data collection and cleaning requirements.

SUMMARY

"Big ideas" require teams with the appropriate knowledge, skills, and experiences to navigate from a "fuzzy" starting point through a myriad of unknown challenges with the hope and desire of turning it into reality. The levers for improving vision delivery lie in architecting the underlying technology infrastructure for high-performance teaming, developing the digital teaming and governing skills to leverage it, reengineering the vision delivery process to create a seamless experience, and inspiring an organizational culture that encourages and rewards collaborative work. The subsequent chapters will do the following:

- Propose the "Digital Office Complex" as an architected collaborative technology environment that enables teams to work together in a unifying manner.
- Illustrate digital teaming and governing practices based on the "Digital Office Complex".
- Introduce a team-centric approach for delivering initiatives, called "Scrumfall".
- Describe potential artificial intelligence (AI) – enabled "Digital Office Complex" capabilities.
- Outline the approach implementation of the "Digital Office Complex" and transform teaming to reengineer vision delivery.

NOTES

1 Vannevar Bush, "As We May Think", *The Atlantic*, July 1945.
2 Mark Sullivan, "'General Magic' Captures the Legendary Apple Offshoot that Foresaw the Mobile Revolution", *Fast Company*, July 26, 2018. www.fastcompany.com/90206157/general-magic-captures-the-legendary-apple-offshoot-that-foresaw-the-mobile-revolution.
3 General Magic Movie, www.generalmagicthemovie.com/.
4 Carl E. Larson and Frank M. LaFasto, *TeamWork: What Must Go Right / What Can Go Wrong*, SAGE Publications, 1989.
5 Michael Hammer, "Reengineering Work: Don't Automate, Obliterate", *Harvard Business Review*, 1990.

Digital Office Complex

2

INTRODUCTION

With ever-increasing competitive pressures, the need to reduce the time-to-value (or time-to-bail) of a "big idea" has never been greater. Figure 2.1 illustrates the conversion flow of a "big idea" – a new product, a new technology, an organizational transformation, a healthcare initiative, or a humanitarian development project – to value by the vision delivery process against a backdrop of volatility, uncertainty, complexity, and ambiguity (VUCA). Teams produce knowledge (capture, creation, sharing), which is expressed as prose, software code, mathematical formulae, and scientific formulae among others, plus decisions that drive the vision delivery process. The ability to improve performance depends upon the quality of knowledge being generated by teams in the vision delivery stream and the speed at which it is shared to support decision-making.

Unfortunately, the current digital infrastructure for vision delivery teams is woefully inadequate. To date, information technology has largely played a supporting role in providing email capabilities to share ideas and feedback among team members and for creating and publishing final work products using personal productivity software. However, this infrastructure impedes better teaming. Email messages and data trapped in documents scattered across the organization work against smooth coordination and collaboration within and between teams as well as making it significantly more difficult to leverage artificial intelligence.

When teams form, they must identify and agree upon objectives. Next, they must create and efficiently execute a plan to achieve them. This requires coordination and collaboration not just among team members, many of whom may be remote, but also between teams. What is needed is an integrated infrastructure that provides team goal management, teamwork coordination, team decision support, team "work product" support, and approval workflow capabilities in addition to "team-to-team" navigation, "team-to-team" coordination, and "team-to-team" data exchange.

DOI: 10.1201/9781003426127-2

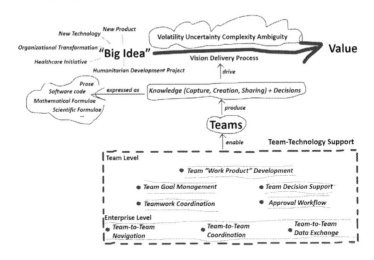

FIGURE 2.1   Converting a "big idea" to value.

The authors define a "Digital Office Complex" ("Complex") as an architected collaborative infrastructure designed to provide this support so teams can work together in a unifying manner. A Complex contains Digital Offices, which are digital workspaces designed for teams and play a role similar to a team's physical space sometimes referred to as a "War Room". A Digital Office is where the team's digital work is visible to all team members and appropriate stakeholders. It is where team members develop their objectives and work plans, update their status, and develop work products collaboratively. Unlike a "War Room", which requires people to be in the same physical location, a Digital Office can support team members and stakeholders working in different geographic locations. Because a Digital Office is part of a Digital Office Complex, teams can share plans, status, and work products transparently.

TEAM SUPPORT NEEDS IN A DIGITAL OFFICE

The following are the five major team needs that a Digital Office must support:

1. Team goal management.
2. Teamwork coordination.
3. Team decision support.

4. Team "work product" support.

5. Approval workflow.

Team Goal Management

As part of team formation, team members must define their purpose and agree on goals. As part of managing their goals, the Digital Office must provide a method for them to identify metrics and track actual results against targeted goals for the team to gauge progress toward them.

Teamwork Coordination

Once the team has clarified its purpose and defined its goals, the team needs to identify actions that various members will take to move them toward achieving their goals. These actions will have target due dates and may have dependencies with other actions. The Digital Office must provide capabilities for the team to develop a schedule of actions and provide team members with the ability to update them so that all members will have access to the real-time status of the team's work plan.

Team Decision Support

A fundamental technique for teams is brainstorming and evaluating ideas, which can be applied to a wide range of problems that teams are tasked to address. In addition, there are technology support needs for more specialized models to apply to specific types of problems, such as stakeholder analysis, Force Field Analysis, and multicriteria decision analysis.

Team "Work Product" Support

While there are specialized forms of work products, such as software, geological maps, and molecular structures, nearly every team will express some portion of their output in the form of documents, such as research findings, initiative requirements, product recommendations, and operating policies. A collaborative document is a capability that enables a group to work on different sections of a document in parallel. They also provide the ability for team members to pose questions, provide responses, and make suggestions for each section of the document. A team notebook is a collection of collaborative documents. These capabilities enable team "work product" development, which is distinguished from "work product" publication. The former is a creative process that focuses on the creation of "content", while the latter is a production process that focuses on the final form of the deliverable.

Collaborative documents and team notebooks are used to support the team in collecting and organizing its research, findings, and thoughts. This aspect of "work product" development is exploratory. The team may start with a high-level structure to guide the development of the "work product". As the team is "thinking through" and developing the content that will become the final deliverable, the collaborative document will grow organically as team members insert new knowledge into the document structure as well as re-arrange it. Some produce "final deliverables", but often is the case that the team notebook is a "living" knowledge repository for team members to reference and continue developing.

Unlike a word processing document that "belongs" to an individual who controls what is added or changed, a collaborative document, or notebook, is owned by the team. The team must develop rules for how information is added or changed and support feedback from each other to refine the content.

Approval Workflow

An approval workflow enables the team to route a "work product" in the form of a document for "sign-off" indicating approval or rejection. For use within the team, it serves as a method for confirming team member agreement or consensus. Examples of using the approval workflow for internal teamwork include all team members "signing-off" on the team charter, team goals, meeting minutes, and decisions. It provides a means for each team member to indicate her agreement. Externally, teams need approval for completed deliverables, changes in the scope of work, and proceeding to the next phase among others. In these cases, the team is seeking "authorization" of what is indicated by the approving item. These "work products" are stored in the "final" form with signatures from the approvers, so they are referenceable at a later time.

DIGITAL OFFICE

A Digital Office is a structured set of collaborative technology components, called Digital Office Objects, that are tailored to facilitate a variety of teamwork. Each Digital Office is designed to support the data and information

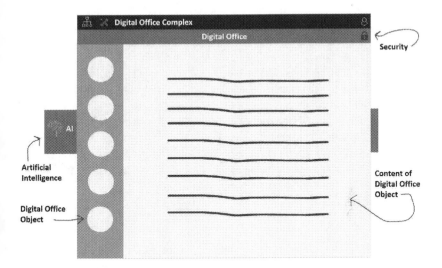

FIGURE 2.2 Digital Office mock-up.

needs of a specific framework, such as strategic performance management, portfolio management, and project management that the team will be using. They comply with standards for user interface and the data structure as part of a Digital Office Complex. The common interface provides a consistent user experience when team members access different Digital Offices within the Complex, while the data structure enables Digital Offices to share data. Figure 2.2 is a mock-up of a Digital Office, highlighting the Digital Office Objects, including their content, artificial intelligence support, and security.

Digital Office Objects

Digital Office Objects are collaborative information structures designed to support the five team support needs described in the previous section. Figure 2.3 illustrates nine basic digital office object constructs that can be used and tailored to support specific team needs.

1. Table – content organized in a row-and-column format that can be displayed in different ways using filters and sort order. Examples: Idea Bank, Stakeholder Register, Risk Register, Issues Log, Meeting Log, and Lessons Learned Log.
2. Outline – content organized in a hierarchical structure. Example: Decomposition of Major Business Processes.

3. Collaborative Documents – a specialized version of the "Outline" Digital Office Object. They facilitate the organization and sharing of the team's knowledge – hypotheses, concepts, findings, learnings, assessments, and member feedback. They are described in more detail in a subsequent section, "Collaborative Documents and Notebooks". Examples: Scientific Research Notes, Legal Research Notes, Vision Development, Policy Development.
4. Discussion – a structure that supports long-format asynchronous discussions. Example: External Environment Scanning Discussions, such as Competitors, Emerging Technologies, and Political News.
5. Schedule – an extension of an Outline Digital Office Objective that provides task dependencies, dates, and costs.
6. Kanban – a specialized display of a table that displays content as a Kanban board. Examples: Activities, Business Proposals, and Work Requests.
7. Measures – a structure for tracking series data to compare actual results against a target. Example: KPIs.
8. Approvals – a structure for managing the approval of documents.
9. Custom – a method for handling custom-built objects. Example: a dashboard for a leadership team that consolidates the status of the schedule, budget, risks, issues, and changes from multiple initiative teams.

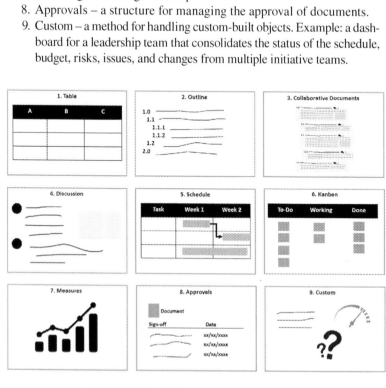

FIGURE 2.3   Nine Digital Office object constructs.

Each record, or item, within a Digital Office Object contains fields to hold content. These fields can be filtered, sorted, and grouped when displayed. They must also comply with Digital Office Complex standards so that data can be shared between Digital Offices and facilitate a consistent user experience.

Additionally, Digital Office Objects require the following capabilities to support and enhance the coordination, collaboration, and communication needs of teams:

- *RACI Support.* A facility to identify and track team members and stakeholders in the roles of "responsible", "accountable", "consulted", and "informed" (RACI) for action-based items or records within a Digital Office Object. Whether the record in a Digital Office Object is a task in a schedule, an item on a Kanban, or a section in a collaborative document, teams identify the one member who is "responsible" for ensuring that it is completed by a specified date. Identifying the team members, or stakeholders, for these roles serves to make clear "who" is expected to complete the specific effort. It is also used as a coordination mechanism to automate notifications, such as due date reminders, status updates, or delays because of unexpected issues to those involved, including those who may have a downstream dependency and a member of a different team.
- *Replies Support.* A facility to enable team members to provide feedback, pose questions, and provide answers to any item, or record, within a Digital Office Object.
- *File Attachment Support.* A facility to attach, or link to, one or more files to any item, or record, within a Digital Office Object.
- *Combine Digital Office Objects.* An ability to combine one or more Digital Office Objects to support complex team information needs.

A Digital Office will contain one or more Digital Office Objects based on these nine constructs, and there may be multiple Digital Office Objects based on the same construct.

Collaborative Documents and Notebooks

A collaborative document is one of the nine Digital Office Object constructs described previously that facilitates the organization, development, and sharing of the team's knowledge – hypotheses, concepts, findings, learnings, assessments, and member feedback on all of them. Figure 2.4 is a collaborative document mock-up.

The gray boxes with outline numbers (1.0, 1.1, 2.0) represent sections within a chapter, and the indentation represents a relationship of sub-sections

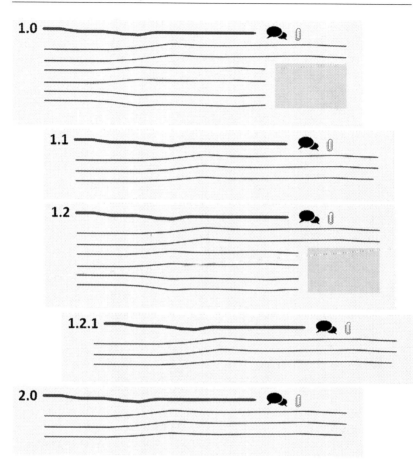

FIGURE 2.4   Collaborative document mock-up.

to a higher-level section. Each section, or sub-section, may contain text, video, images, and links to external sources. Additionally, the "discussion" bubble and "paperclip" icons provide team members with the ability to share comments and attach files to the section.

This collaborative document construct lets the team's "knowledge" grow organically because team members can add to the content at any time. New sections, or sub-sections, can be created on-the-fly, moved around, or combined with others. It is not an individual's notebook; rather, it is the team's notebook and represents the collective work of the team. Team members can add content to any of these items at any time. It is not a personal document that is sent from one person to another or checked in/out of a document library before a team member can edit it. Multiple sections can be worked on at the same time by different team members.

Grouping collaborative documents simulates a physical "notebook" organized into chapters, sections, and sub-sections that team members can review and contribute to at any time. It is a team "notebook". Example uses of team "notebooks" are as follows:

- Legal team working on a complex case. Team members organize the team "notebook" to document their research and discovery work for a specific case.
- Team preparing a grant proposal to fund a new research initiative.
- Team organizing a conference with break-out sessions organized around tracks.
- A team organizing the features of a new product.
- A cross-disciplinary team of researchers from multiple universities developing hypotheses to test and capture results for a "grand challenge".

Artificial Intelligence (AI)

Artificial intelligence (AI) assists and augments team intelligence for many different purposes. For one, the rich content stored in a Digital Office forms a "structured knowledge repository". AI provides a capability to analyze and summarize a "snapshot" for teams to use as part of evaluating their performance and determining the next steps. For example, an initiative team may ask, "Assessing our current performance, what processes might be employed to improve the outcome?" Alternatively, an executive might ask such questions as "What are the highlights of last week's progress regarding our digital transformation program? What are the top concerns? Are there changes should we consider?" AI is covered in more detail in Chapter 6.

Security

Security determines who has access to team data and what level of access, such as read-only, read and contribute, edit, and grant-and-revoke access. Each team determines who has permission to access its Digital Office. It is also concerned with exceptions, such as a person who does not have permission to access the team's Digital Office but may be granted permission to access and contribute to a specific item in a digital office object, such as an "issue" that has been identified that involves assistance from an external party.

VISION DELIVERY MODEL

The vision delivery model describes the vision, strategy, portfolio, and initiative processes specific to an organization. Figure 2.5 illustrates a general vision delivery model. Each layer – vision, strategy, portfolio, and initiatives – has its unique data structure. Its purpose is to inform the construction of Digital Offices to support teams at each layer in addition to the data exchange needs between layers. The high-level process begins at the vision layer and outputs flow downward to become inputs through the strategy, portfolio, and initiative layers before returning upward through the same layers in reverse. With each process cycle, the goal is to learn more that can be used in developing assets and preparing for implementation and launch.

Layer 1 – Vision

While there is no single approach to visioning, the ultimate purpose is to express the "big idea" in an explicit form. The process may start with a small team that describes the essence of the "big idea" through storytelling, visuals, or a prototype. The team may conduct workshops with different stakeholder groups to get their perspectives and opinions. In addition to the illustrations that convey the essence of the "big idea", the team will capture narratives that

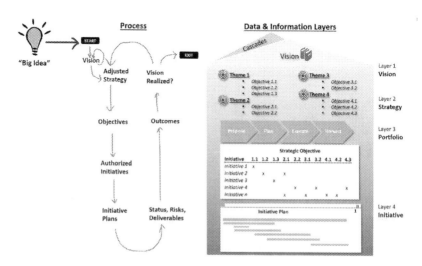

FIGURE 2.5   Vision delivery model.

detail how things will be different as a result of its implementation because the "big idea" represents change, which likely will be profound and will require consideration when developing the strategy. Once the visioning elements have been sufficiently defined, the strategy work can begin though visioning work continues.

Layer 2 – Strategy

The leadership team uses the "vision" as input to formulate the business strategy to deliver on the vision. It outlines "what" objectives need to be achieved, "when" they need to be achieved, and "what" measures will be used to evaluate progress to determine whether it has been achieved or not. These objectives are typically organized into key themes, such as "Financials", "Customers", "Internal Processes", and "Organizational Growth and Learning" to emphasize the most significant areas of focus for the organization. During each process cycle through the layers, the leadership team reviews the performance evaluation and may need to make course corrections to the strategy before proceeding.

Layer 3 – Portfolio

The strategy informs the portfolio team as they evaluate initiative proposals to make investment decisions. These proposals are scored based on criteria, such as alignment with the strategic objectives, risk level, investment and resource requirements, and benefits. An overall score is used for selecting, prioritizing, and scheduling initiatives. The portfolio schedule lays out the road map of work to be done relative to the strategy. The authorization of the initiatives moves to the "initiative" layer for a team to create an Initiative Plan and execute it. The portfolio is continuously monitored for the progress of and the benefits from the initiatives.

Layer 4 – Initiative

The input to the initiative layer is the authorization to form initiative teams to plan and execute the work necessary to achieve the specific objectives outlined in the strategy. At regular intervals, these teams provide the status of their work, budget, and risks to the portfolio team to analyze and summarize for review by the leadership team who assess in the context of their strategy, current business conditions, and external environmental factors.

Cascades

With larger organizations, the strategy cascades downward to business units, divisions, departments, functions, agencies, or ministries. Each unit uses this same vision delivery model. Each unit's vision and strategy are aligned with the corporate vision and strategy.

DIGITAL OFFICE COMPLEX TO ENABLE VISION DELIVERY

A Digital Office Complex ("Complex") contains two or more Digital Offices, one of which supports the leadership team with organizing the vision, strategy, and portfolio of initiatives. A Complex itself may contain Complexes to further organize Digital Offices for business units, departments, or divisions. Collectively, these Digital Offices that comprise a Complex form a "structured knowledge repository". Figure 2.6 is a mock-up of a Digital Office Complex with the following highlights:

1. Leadership Digital Office plus Digital Office Complexes for two business units.
2. Multiple Digital Offices within each "Complex".
3. Digital Office Objects with field content within each Digital Office.
4. "Team tools" include digital brainstorming and team decision support capabilities.
5. The Organizational Directory maintains a record with an address, or pointer, to digital offices and their relationships to other digital offices within the digital office complex.
6. User profile defines user-configured preferences to guide each user's experience.

In the mock-up, the Leadership Digital Office has a dashboard display that contains the organization's vision and strategic objectives. The frames labeled "teamwork", "business outcomes", "risks and issues", and "radar" are driven directly by data and content from the Digital Offices within the Complex. The appropriate data and content automatically transfer from the individual Digital Offices to the Leadership Digital Office. It does not require each team to extract the requested data and email it to a contact on

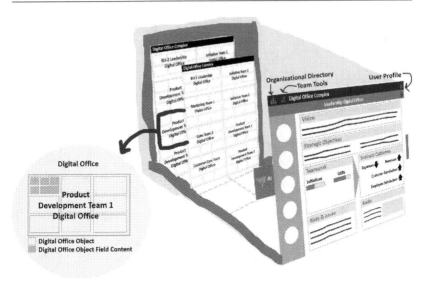

FIGURE 2.6   Digital Office Complex mock-up.

the leadership team. "BU-1 Digital Office Complex" contains Digital Offices for three product development teams, an initiative team, a marketing team, a sales team, and a customer care team with a similar set for the "BU-2 Complex".

Other relevant information that provides context for performance are the "risks and issues" and "radar" frames. Major risks and issues that are likely to have profound impacts on performance bubble up automatically from the teams. The "radar" Digital Office Object provides news and analysis regarding the external environment that may impact the organization's strategy. Proactive leaders can navigate to any team's digital office not to micromanage but to stay abreast of the work rhythm and key issues for which they may be able to assist, such as making a connection between a member of one team to an expert in a completely different organization.

Team Tools

Team decision support plus communication tools, such as messaging, videoconferencing, and whiteboarding, comprise the category of common team tools in the digital office complex architecture. Outputs from these tools can be stored in a Digital Office.

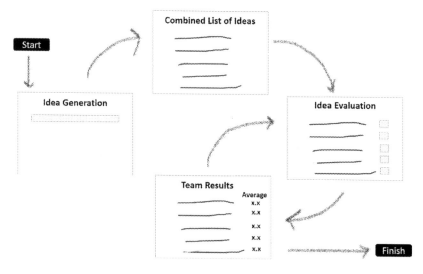

FIGURE 2.7   Digital brainstorming and team decision support mock-up.

Team decision support capabilities include idea generation, categorization, and evaluation. Three components comprise the foundation for digital team problem-solving:

1. Idea generation (digital brainstorming). Capability for individual team members to enter ideas anonymously.
2. Idea categorization. Capability to group ideas into categories that can be evaluated.
3. Idea evaluation. Capability for individual team members to rate ideas anonymously. Rating schemes include multipoint scales (e.g., 5-point scale), points allocation, and multidimensional (e.g., 2 × 2 matrix).

Figure 2.7 is a mock-up of digital brainstorming and team decision support capabilities. Each team member has a screen to enter "ideas" anonymously. It will display the combined list of ideas from all team members. Various techniques can be used to evaluate the collection of ideas, such as a 1–5 rating scale and fixed-point allocation. After team members have evaluated the ideas, the results are displayed for the team to discuss after which the team may decide to re-evaluate the list.

Research has shown the following impacts of digital brainstorming and team decision-support capabilities:

- Team members' interest and involvement during a meeting increase.
- Individuals who are otherwise quiet participate.

- Individuals whose ideas are discounted by others when presented verbally get more recognition for their ideas.
- Team members leave the meeting with something concrete to think about for the next meeting.
- When brainstorming and evaluation have been used for specific issues, people leave the meeting with clear ideas and priorities.

In the late 1980s, as part of their research to enhance the effectiveness and efficiency of face-to-face meetings, researchers Geradine DeSanctis, Scott Poole, and Gary Dickson at the University of Minnesota created the SAMM (Software Aided Meeting Management) system. SAMM provided idea generation, idea evaluation, and decision aids to support problem-solving teams.

An information technology research team at Texaco led by the authors partnered with the UM researchers to explore the potential use of SAMM to support their teams that were newly trained in Total Quality Management (TQM) processes and techniques. The Texaco team turned a 330-square-foot storage room into a "Decision Room" with a custom-built, U-shaped table, ten monitors recessed to not obstruct a team member's view, a projection screen in the front, and a server to run SAMM. Throughout their three-year study, there were over 30 teams that used it. Some teams held their weekly meetings there whether they used SAMM or not. Many teams used it for a one-time, specific purpose, such as a strategy session, vendor selection, and JAD (joint application development) sessions.

The expectations of SAMM in supporting these teams included full and even participation in discussions, idea generation, democratic choice making, and orderly, documented meeting processes. The teams used SAMM to display agendas and record information that provided continuity in work from one meeting to the next. When SAMM was used for brainstorming, high-idea generation and participation by all members occurred. In addition, decision quality improved when SAMM was used, in the sense that creative ideas were put forth and the teams engaged in more thorough problem analysis. One team, whose charter was to automate the operations of the data center, used SAMM to brainstorm on the topic of "When is the CPU down?" Team members simultaneously entered responses to the question. At the end of the brainstorming activity, the combined list of ideas was displayed on the group screen in the front of the room. In SAMM, the ideas were displayed anonymously. By not having a team member's name associated with the idea, the theory was that the team could focus on the idea rather than the contributor of the idea. Following the display and discussion of all the ideas, the team realized that the answer to their question depended upon the perspective of the person answering the question: software, operations, or tuning.

There were several notable impacts of SAMM. For one, the teams used their TQM processes. They planned for their team meetings. Before a meeting, they thought through what it was they wanted to accomplish, what processes they would use, and how SAMM would be used to support it. One team leader reported that when they could not hold their team meetings in the "Decision Room" and, therefore, did not have access to SAMM, team members would say, "Let's do it the SAMM way". By using SAMM to collect ideas, all team members generated ideas simultaneously, which provided fairer participation in the process. Without SAMM, it was not uncommon for one or two members to dominate the meetings because they took a great deal of "airtime" or because shy team members wouldn't say anything.[1]

Organizational Directory

The Organizational Directory organizes digital offices into a hierarchy that assists in user navigation as well as automated dashboarding and reporting. The hierarchy provides a structure for groupings, such as business units, functions, departments, and divisions.

User Profile

The content within a digital office grows during the team's life. A person who is a member of one team often has dependencies with the activities of another team. As team members update content in the various digital office objects, the ability for a user to be made aware of information that is the most relevant and urgent, quickly find the content, and assess it within the context of the team's work is a critical need. The user profile stores a custom configuration for each user's experience. The profile may include the ability for users to customize the types of notifications they receive, such as changes made to items within digital office objects, reminders about an item with an upcoming due date, or a newly documented issue on a dependent activity.

USE CASE: DIGITAL OFFICE COMPLEX FOR A POST-MERGER INTEGRATION INITIATIVE

The vision for the merger of Company A and Company B is "a more efficient company, NewCo, that will bring the best products from both companies plus innovative new products to its customers". To realize this vision, six leadership teams were formed comprised of people from both companies. These leadership teams focused on six business areas: marketing, manufacturing, finance, legal, purchasing, and information technology. Each leadership team formed the number of teams it needed to recommend a new organization for its respective business area. The information technology leadership team created three teams – infrastructure, information systems, and data – to "figure out" how to provide the most efficient and effective digital infrastructure to support "NewCo" out of the information technology assets from the two companies.

The mock-up of the Digital Office Complex to support the entire integration organization for all six business areas is shown in Figure 2.8. It consists of six Digital Office Complexes, one for each business area, within the "NewCo" Digital Office Complex. This mock-up highlights the information technology Digital Office Complex.

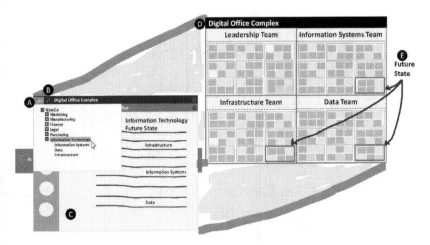

FIGURE 2.8 Mock-up of Digital Office Complex for a post-merger initiative.

A. Organizational Directory

The Organizational Directory provides a hierarchical structure for all six business areas – marketing, manufacturing, finance, legal, purchasing, and information technology – and the teams within each of these areas. It is a navigational tool to assist a user in accessing any Digital Office in the organization. Team members from one team are very likely to have dependencies with one or more of the other teams. Being able to navigate to the Digital Offices for those teams allows a team member to scan and stay abreast of relevant issues, decisions, and timing. For example, by having visibility into the marketing teams' discussions regarding moving employees out of their current office space and into a new office building, the infrastructure team members can anticipate the requirements to support the new office building and the work to relocate physical infrastructure assets to the new location.

It also serves as a relationship map for reporting purposes. To create a consolidated report for the information technology leadership team, the Organizational Directory has the references to the three Digital Offices – infrastructure, information systems, and data – which is needed to locate and access the relevant data to produce a report.

B. Team Tools

Digital brainstorming and team decision support capabilities are among the team tools which is important for team activities, such as brainstorming and prioritizing implementation risks, stakeholders, and success criteria.

C. Digital Office User Interface

While a Digital Office may be tailored for each team, as part of a Digital Office Complex, each Digital Office is organized with a standard set of Digital Office Objects to facilitate the navigation from one to another. The mock-up illustrates a consolidated report for the "Future State" of information technology that consolidates the work from its three teams.

D. Digital Office Complex Within a Digital Office Complex

The Digital Office Complex consists of Digital Offices that are addressable programmatically. Each Digital Office is a container of Digital Office Objects including their data. A Digital Office Complex may contain other Digital

Office Complexes. To support the "NewCo" merger integration endeavor is the "NewCo" Digital Office Complex comprised of six Digital Office Complexes – one for each business area. As shown, the information technology Digital Office Complex is comprised of four digital offices, one for the information technology leadership team and one each for three sub-teams – the infrastructure team, the information systems team, and the data team. In this mock-up, each Digital Office contains nine Digital Office Objects that have been configured with the same fields to facilitate automation as well as user navigation.

E. "Future State" Digital Office Object

In the mock-up, the "Future State" Digital Office Object is outlined for three teams. Because the "Future State" Digital Office Object is addressable programmatically, a report, which is shown on the information technology Digital Office, can automatically retrieve the data from the "Future State" Digital Office Objects from each of the three teams. It does not require each team to extract the content, publish it in a word processing document, and email it to the Leadership team.

SUMMARY

With ever-increasing competitive pressures, the need to accelerate the process of turning "big ideas" into reality has never been greater. Much of the current implementation of information technology to support organizations is a patchwork of technology and data, or "Islands of Digital Teams", that impedes teams from working together which, in turn, impacts performance. This chapter covered an architectural design for a Digital Office Complex built on database technology to enable data sharing within and between teams. Improvement opportunities lie in the reengineering of vision delivery through the implementation of a Digital Office Complex and transforming teaming. Among the improvements to result from replacing the "Islands of Digital Teams" with a Digital Office Complex include the following:

- Simplification of the digital infrastructure.
- Integrated and seamless user experience.
- Flattened communication hierarchy that provides leaders a means to directly access any team's Digital Office. This direct access replaces sending messages up and down the hierarchical management layers.

- "Single source of truth".
- Elimination of administrative time spent extracting and reformatting data from word processing and spreadsheet files from one team to another for consolidated status reporting.
- Data captured in one place and used broadly by many.
- Ability to share "lessons learned" in real-time between teams within a large strategic initiative. As teams begin implementing, they are very likely to encounter unanticipated problems. How they resolved them can be shared through "lessons learned" that other teams executing at the same time can avoid.
- Real-time alerts regarding issues with dependent tasks between teams.
- Tighter coordination and collaboration between teams.

Digital Office Complexes are assets to be managed and leveraged similarly to other assets. The by-product of teams working in Digital Office Complexes is data, which is used to learn and improve organizational performance. Much as physical office space is designed for physical teamwork, digital office complexes enable teams to work in a unifying manner coordinating their efforts to produce outputs in a timely fashion. Unlike physical office space, though, Digital Office Complexes support teams whether team members are in the same physical location or remote.

NOTE

1 Gerardine DeSanctis and Brad Jackson, "Managing the Team-Based Organization: Perspectives from Texaco", *Presented to MIT Center for Coordination Science*, 1995.

Digital Teaming 3

INTRODUCTION

The authors define "digital teaming" as team members working together and with other teams using a Digital Office Complex. A team's Digital Office supports them with capabilities described in Chapter 2: Team goal management, teamwork coordination, team decision support, team "work product" support, and approval workflow. Team members must acquire new skills and team practices to achieve the benefits of digital office capabilities. The following sections describe digital teaming techniques supported by capabilities within the digital office complex.

DIVERGENT/CONVERGENT THINKING AND DIGITAL BRAINSTORMING

Divergent/convergent thinking is a core tenant of digital teaming. This concept is inspired by psychologist J. P. Guilford's work in the 1950s.[1] At the beginning of the process, the team brainstorms as many ideas as they can. They are "expanding the list of possibilities", which represents what Guilford calls "divergent thinking" followed by "narrowing the possibilities", or convergent thinking as illustrated in Figure 3.1.

Using digital brainstorming, often in less than a minute, the team will have a large list of ideas for the topic they are undertaking, such as ideas for features of a new product. Because each member brainstorms privately and anonymously, there will likely be duplicates in the list. The next step involves removing duplicates. Some ideas on the list will be exact duplicates while others will be similar, and the team will need to discuss whether it should be considered a duplicate and deleted or retained for further consideration. This step will reduce the list somewhat. The next step involves team members evaluating

DOI: 10.1201/9781003426127-3

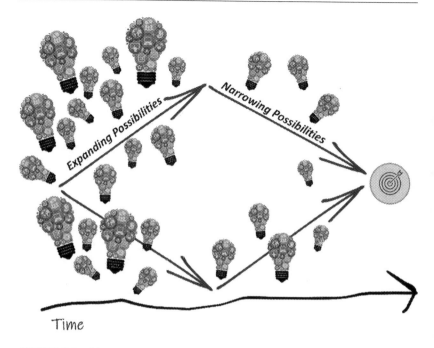

Time

FIGURE 3.1 Divergent/convergent thinking model.

the list of ideas. There are different evaluation techniques, such as points allocation, and rating, that the team may use.

Digital brainstorming is more than just an automated version of its manual counterpart. The benefits include greater and more even participation, creativity, and improved decision-making. With digital brainstorming, team members generate ideas anonymously and simultaneously. It allows everyone's ideas to be included with a larger number of ideas being generated, which generally leads to more creative solutions. Increased involvement by team members leads to a greater sense of team, rather than individual, ownership of the ideas generated.

Evaluating ideas aids convergent thinking. Two common evaluation techniques are "points allocation" and rating. The "points allocation" uses a fixed set of points that are given to each team member to allocate across a list of ideas. They can allocate all the points to a single idea or distribute varying amounts across multiple ideas. If the pre-determined fixed number of points is "5", a team member may allocate "5" to a single idea, "1" to five different ideas, or "2" to one idea and "3" to another, and so forth. The total points given to all ideas cannot exceed "5". The manual method for this technique is to provide "sticky dots" that team members use to post next to the idea written on a "flip chart". The ideas receiving the most "sticky dots" are the "top" ideas.

In the digital version of this method, team members evaluate privately and anonymously, whereas in the manual method, team members may be influenced because they can see where other team members are placing the "sticky dots" and how many they place. This method is useful when the team has a long list of ideas. One advantage is that team members do not have to assign a value to every idea as they do with the "ratings" method, which means they do not have to evaluate every idea just their top ones.

The idea evaluation phase begins an iterative process of reducing the larger list of ideas based on the opinions of team members. Digital evaluation not only provides a quick calculation of the team average rating for each idea but also the range and extent to which team members agree with the average rating. This additional information helps to point the direction of what needs to be discussed to gain and ensure clarity for the most important ideas.

The rating averages represent the team's collective evaluation. The extremes for each idea are presented as "minimum" and "maximum". If there is a big gap between these, then at least two members are at opposite points in terms of viewpoint on that idea. If the evaluated list is important to the team's goal, and consensus is important, then team members need to discuss those with disparate scores. Often, teams choose to re-evaluate ideas after discussing them to see if the members are closer to consensus, or at least in the same ballpark.

An average rating of "5", on a 5-point scale with a standard deviation of "0" indicates it is a top idea and that all team members agree that it is. In contrast, an idea with an average rating of "3" with a min of "2" and a max of "5" and a standard deviation of "1.45" indicates that there is disagreement among team members about the idea. At least one team member rated it very low while another thought it was an excellent idea. With this additional information, the team can explore why it would rate in the extremes.

DIGITAL TEAMING DEVELOPMENT STAGES

Teams use their Digital Office throughout the team's life as illustrated in Figure 3.2.

- Stage 1 – Team formation is the identification of team members who will work together to produce the outcomes for which they have been charged. Digital Office Organization is the identification of the digital office objects needed to support the team and interconnect with other teams within the digital office complex.

FIGURE 3.2 Digital teaming development stages.

- Stage 2 – Team goal setting and pathing starts with the team understanding their purpose, setting goals, and defining a pathway to achieve them.
- Stage 3 – Team delivery involves the team defining and coordinating their actions to produce the necessary outcomes.
- Stage 4 – Continue or exit? If the team has fulfilled its purpose, it disbands. Otherwise, it continues onward to bring value.

STAGE 1 – TEAM FORMATION

As team members are selected to form the team, most of the activity in this stage is "preparing the team for success". It includes configuring a digital office, which will become part of the digital office complex, for its specific teamwork. It involves team building, clarifying its purpose, and gaining commitment to it.

SCENARIO – OPEN MEETING AGENDA

"Appears everyone has gotten all their topics in. Unless anyone needs more time, we'll look at all the ideas now", announced Courtney after 30 seconds of digital brainstorming. Courtney, the lead for the firm's new robotic consulting practice, starts the meeting by asking her leadership team members to enter the top items on their minds. Members' ideas are anonymous, and they can't see each other's while brainstorming. She then displays the full list, which is 35 topics. With a team of seven, that's five topics per person in 30 seconds. Morgan, the newest manager to join the practice, comments that the topic, "Hiring" appears several times. Fellow team member Preston wonders aloud if the items "Hiring" and "Resources" mean the same thing. After a few minutes of discussion among all the team members, Hannah deletes the duplicate "Hiring" items but leaves "Resources", as they all agreed it had a different meaning from "Hiring". The list has been reduced to 12 items, which was still too many to discuss in the hour time that they had. Courtney then guides the team to evaluate these 12 items that are potential "agenda topics" so they can reduce and prioritize the list. "I'm giving each person five points to allocate toward the topics that you feel are the most urgent to discuss. You can allocate all to one topic or across multiple". In less than 30 seconds, all team members were finished. Courtney displayed the results on the screen in the front of the room. "Looks like we're talking about 'Hiring' as the first agenda item, as it had 17 points followed by 'How to arrange workspaces' second, which received seven points. We'll keep this list to cover in subsequent meetings". As this was Morgan's first team meeting at this firm, she contrasted this experience to her previous firm, "That was an amazingly fast way to create an agenda with all of our input and organized with the most important ones first".

Meetings come in all shapes and sizes. Some are well-planned with very specific agenda topics, while others are organized on-the-fly but with purpose. The scenario described earlier depicts a type of meeting that is not preplanned but organized at the start of the meeting based on what's the most interesting and urgent to the members. It is made possible by digital brainstorming and team decision-support capabilities. It can take less than five minutes to collect ideas for agenda topics, rate them in terms of importance and urgency, and discuss ones that require clarification. Figure 3.3 illustrates a team member distributing a fixed set of points to the team brainstormed list of meeting topics to indicate those of the most interest and importance to him.

Once all team members complete their voting, the results are shown as illustrated in Figure 3.4. The remainder of the meeting time is spent focusing on those items with the most value to the team members.

In the early stages of team formation, hearing from all team members as to their top concerns and the team working through those topics is a very

FIGURE 3.3   Team member distributes ten points across potential agenda topics.

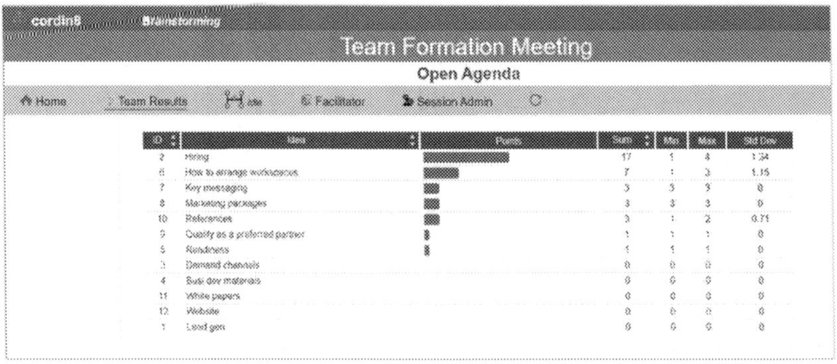

FIGURE 3.4   Team results from voting for agenda topics.

powerful team-building exercise. Storing the outcomes of the meeting, including the agenda, minutes, action items, and decisions in the team's Digital Office retains it as part of the team's memory.

STAGE 2 – GOAL SETTING

Having formed the team and achieved clarity regarding its purpose, the team is ready to identify its goals. Different methods for setting goals, such as OKRs (objectives and key results) encourage a "bottom-up" alignment of goals from teams to the longer-term corporate goals. Teams can use digital brainstorming and evaluation to collect and prioritize their goals. Along with the goals, the team identifies metrics for periodic evaluation of their performance and progress toward accomplishing them. As a final step in the goal-setting process, the team can route the goals through the approval workflow for sign-off by all team members which contributes to the commitment of each team member to work to achieve their goals.

SCENARIO – CREATING OBJECTIVES

"I'd like some discussion on 'Make *Place* No Longer Relevant' as an objective", commented Tanner after the results were displayed on the screen in the front of the room. "The average rating is a '4.1' on the 5-point scale, which is high. I rated it as a '2'. While technology does remove the barriers associated with being able to work together remotely, at the same time, it makes the 'place' where one works – home office, coworking office, city, state, country – crazy relevant. It allows an employee to make two separate decisions – 'Where do I want to work?' and 'Where do I want to live?' instead of whatever company I decide to work for determines where I'll be living. To me, that's incredibly relevant". Fellow team member Josiah who had rated it a "5", explained that he evaluated it as a technology capability focus but now sees it in a different light. Fellow team member Anna interjected, "It reminds me of a recent survey I saw which indicated that employees are split between working at the office and working from home. I rated it a '4', but I think Tanner makes some excellent points. What if we modify the objective to 'Make *Place* a Choice'? We want to create technologies that enable our customers the ability to provide options for 'working'. People like options". After some further discussion, the team agreed to restate the objective as Anna had proposed and then re-evaluated it to see if they had moved toward a consensus on their objective.

The benefit of the discussion is to uncover more details about the idea that can clarify the meaning or intent behind it. It is very often the case that members have different understandings of the ideas. Through discussion, it becomes clearer as to exactly what is meant. Because the calculations are fast, it is easy for team members to re-rate the ideas based on their new insights to see if the team moves toward a consensus on the various ideas.

STAGE 3 – TEAM DELIVERY

Having completed the previous stage, the team is now ready to plan and produce the outcomes that will contribute to achieving its goals. Three significant digital teaming capabilities are as follows:

1. Action plan for teamwork coordination.
2. Team notebooks to support "work product" development.
3. Team decision support.

Action Plan for Teamwork Coordination

A team's "action plan" scales to the size of the endeavor. Less complex undertakings require less planning and can be supported by a team task list, Kanban board, or shared calendar. Those with greater complexity and higher risk require a more robust approach.

Whether an endeavor is small or large, the action plan must be visible to all team members and stakeholders. Team members must keep it updated with interim progress, current issues, and interim work products. Team members participating in the development of the action plan is critical as is using RACI, which stands for "responsible", "accountable/approver", "consulted", and "informed", on the "action plan" digital forms. It enables those identified to be notified of updates to an action. Keeping team members updated on the interim status and issues that cause delays is critical. It is part of a fast communication system in high-performance teaming.

Collaborative Notebooks to Support Team "Work Product" Development

Chartering teams to research, discover, explore, or create something "new" is one of the most significant reasons to form a team. An early step in the team's work often involves the team "learning" by building a body of knowledge based on team members' knowledge plus additional research. A team tasked with developing a new policy may start by researching examples of similar work done within the organization as well as externally. This knowledge can be organized in a collaborative notebook for the team. Along the way, the team will begin developing "work products", or deliverables, using collaborative

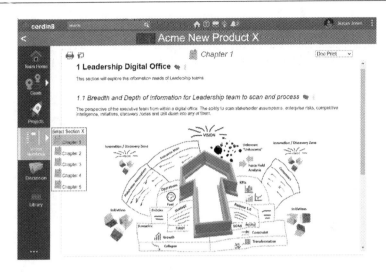

FIGURE 3.5   Example of team's Vision Notebook.

notebooks. They can outline the key themes and messages for team members to use to build-out the completed work product. Any team member can add to it, comment on any of it, and reorganize it. It's a team, shared notebook rather than an individual notebook or collection of individual notebooks. Figure 3.5 illustrates a "Vision Notebook" as an example of a team notebook. It is organized into five chapters displayed on the selection menu with Chapter 1 being highlighted. Section 1.1 titled "Breadth and depth of information for leadership team to scan and process" contains text and images as part of the content. It is one of many entries in the collaborative document. Team members can post questions, provide feedback, and describe alternatives, which other team members may weigh in, to the content in the section using the "discussion bubble" icon. Lengthy discussion threads with unresolved issues should be placed on the team's meeting agenda to resolve. Members may attach files to a section, such as supporting materials, by using the "paperclip" icon. It is an example of how a team can organize the development of their work products to express their thinking.

Team Decision Support

The following are common methods and decision models supported by digital brainstorming that teams can use to support their work.

General Team
- Snow card technique – clustering ideas.

Product Development Team
- MoSCoW technique – prioritizing scope.
- Fibonacci sequence – estimating relative effort.

Problem-Solving Team
- Problem formulation method – problem-solving.
- Fishbone – root cause analysis.

Selection Team
- Multi-criteria decision analysis – selecting among a set of options.

Snow Card Technique

The snow card technique is a "bottom-up" technique where team members generate a broad range of ideas and then group similar ideas into categories for which they label, or name. Then the team works with the "categories" for further discussion and evaluation. Figure 3.6 illustrates the results of a team brainstorming ideas for features of a new product. Through discussion, the team grouped related ideas into categories, e.g., Feature A, Feature B, Feature C, and so forth. The team can now work with the "categories", such as rate them for prioritization.

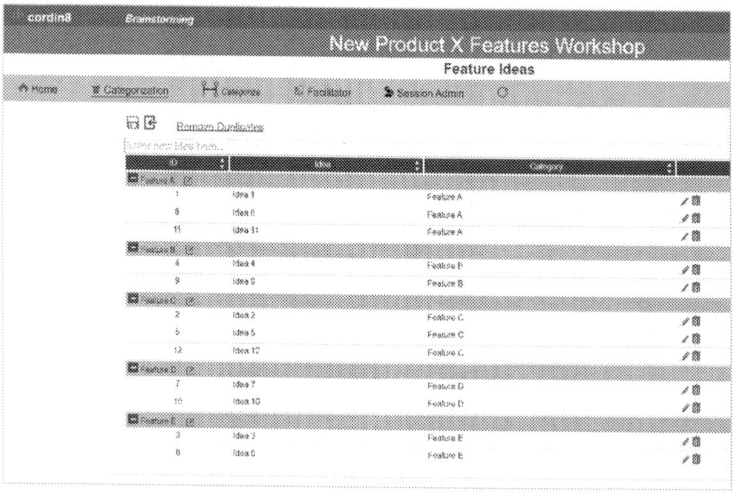

FIGURE 3.6 Grouping similar ideas (features) into categories.

MoSCoW Technique

"Wow! We generated 163 ideas", summarized Kevin, product manager, following a digital brainstorming session to identify features for a new product. "To be able to organize this list, we need to cluster ideas that are related into categories", fellow team member Carolyn described as the next step. After removing 41 duplicate ideas, the team grouped the remaining 122 into 14 categories. Kevin then asked, "Using the evaluation feature, I'd like each person to rate these categories by 'Must Have', 'Should Have', 'Could Have', and 'Won't Have'. That will provide us with a prioritization of the new features".

An illustration of this method is a workshop with participants from both marketing and development organizations. The objective of the workshop is to prioritize a list of feature ideas to be developed. The list of brainstormed ideas is presented on a screen to each attendee. Those from marketing are asked to rate each idea on a 10-point scale based on the value it brings to the customer. A "10" indicates that it brings the most value. The developers are asked to rate the same ideas using the same scale though they are asked how simple it will be to implement. A "10" is the simplest.

Figure 3.7 shows the results of both groups' ratings. The ideas are listed with two sets of results. The first set is the average ratings from the marketing attendees ("Value to Customer"), and the second is the average ratings from the developers ("Simplicity of Development"). The usefulness of this table is to review the ideas where there is significant disagreement among team members. In terms of feature value, the ratings for "Feature G" (ID 8) listed near the bottom of the table ranged from "2" to "8". Similarly, there is disagreement

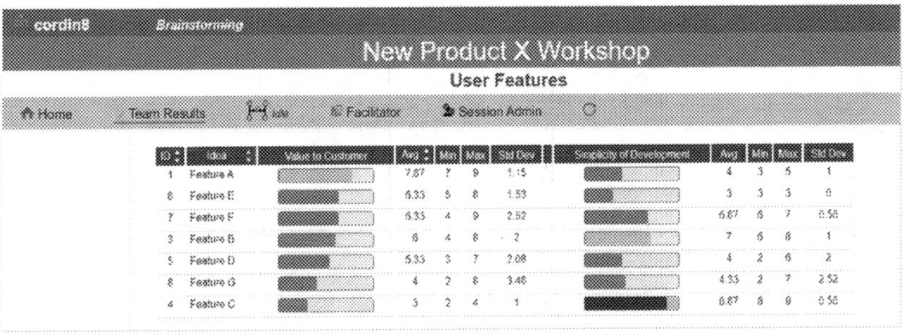

ID	Idea	Value to Customer	Avg	Min	Max	Std Dev	Simplicity of Development	Avg	Min	Max	Std Dev
1	Feature A		7.87	7	9	1.15		4	3	5	1
6	Feature E		6.33	5	8	1.33		3	3	3	0
7	Feature F		6.33	4	9	2.32		6.67	6	7	0.58
3	Feature B		6	4	8	2		7	6	8	1
5	Feature D		5.33	3	7	2.08		4	2	6	2
8	Feature G		4	2	8	3.46		4.33	2	7	2.52
4	Feature C		3	2	4	1		6.67	6	8	0.58

FIGURE 3.7   Product features grouped by priority based on team ratings.

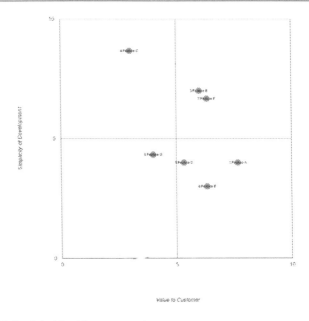

FIGURE 3.8   Prioritized feature matrix.

among the developers for how easy it will be to produce with a range of "2" to "7". Discussions on this idea should lead to a better understanding of the feature and can help them move toward consensus concerning its value and development complexity.

The mapping of these averages into a 2 × 2 matrix is shown in Figure 3.8. Appearing in the upper right quadrant, "Must Have", are Feature B (ID 3) and Feature F (ID 7), as they rated high on both the "value" and "simplicity" dimensions based on the group's opinion. They have the most value to the customer and are the easiest to develop while Feature G (ID 8) had low ratings on both dimensions, so it appears in the lower left quadrant, "Won't Have".

Fibonacci Sequence for Prioritizing Relative Effort

The Fibonacci sequence is a set of integers, which are called the Fibonacci numbers, starting with "0" and "1" followed by numbers that are the sum of the preceding two numbers. The first ten numbers in the sequence are "0, 1, 1, 2, 3, 5, 8, 13, 21, 34". Using a modified version of this sequence is a common approach to estimating the effort of a complex task. Instead of guessing how much effort a task might take, the team assigns a Fibonacci number based on relative comparison to other tasks. For purposes of estimating tasks, the sequence starts at "1" instead of "0", and some teams make further modifications by using "20" and

"30" instead of "21" and "34". A task that is rated a "1" is assessed as one that requires the least effort with a high degree of certainty for its completion. One that is rated a "2" suggests it will take twice as long as a "1" with slightly less degree of certainty. As the numbers get higher, the task is considered more complex, will require more time to complete, and has a greater degree of uncertainty.

Using digital idea evaluation supports team members in arriving at an agreed-upon rating or estimate. Software development teams often use an estimating unit called "Story Points". Figure 3.9 shows a team member evaluating a list of "stories" using values from the Fibonacci sequence.

Once all team members have completed their evaluations, the average ratings are shown as depicted in Figure 3.10. The first two ideas listed, "Story 2"

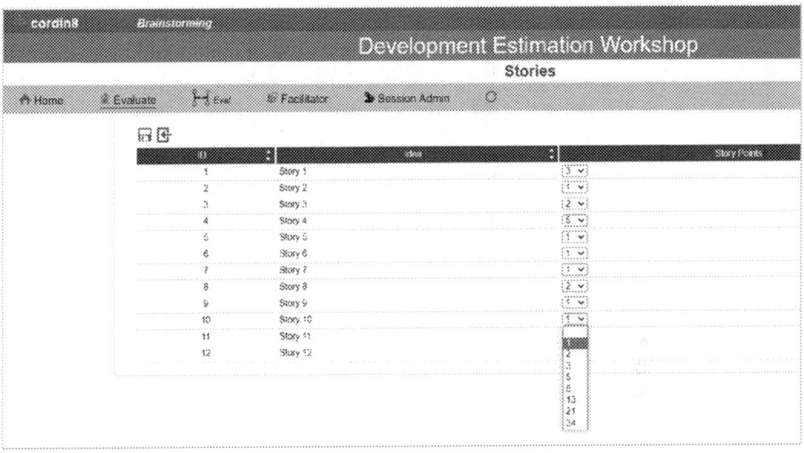

FIGURE 3.9 Team member estimates the number of Story Points for each idea.

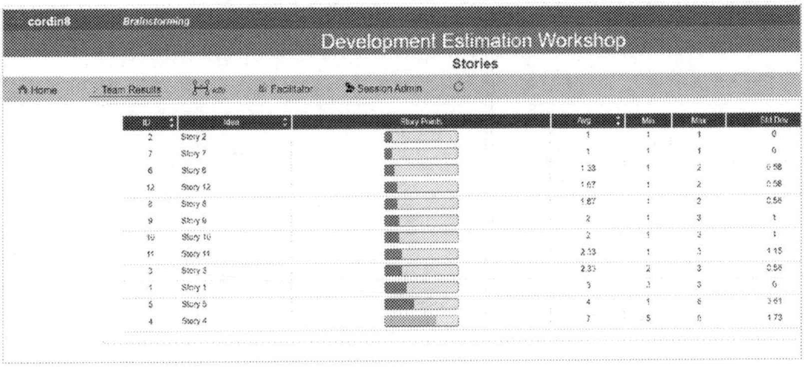

FIGURE 3.10 Average Story Point estimation for each story listed from lowest to highest.

and "Story 7", have an average rating of "1" for each. With a standard deviation of "0", there is complete agreement that these two ideas will be easy to develop. On the other hand, "Story 5" listed next to last in the table shows a range of "1" to "8" which means someone believes it is simple to do while at least one person believes it is eight times more difficult. The team needs to discuss why there is such a discrepancy.

The table also sorts these ideas, or "stories", from the easiest to the most difficult, which may serve as the prioritized order.

Fishbone

The fishbone technique, which is also known as Ishikawa, is a team problem-solving method that categorizes cause and effect. It was popularized in the 1960s as a basic tool of quality control by Kaoru Ishikawa at the University of Tokyo. It takes its name from the resulting diagram that resembles a fish skeleton laid on its side. The problem is written at the mouth of the fish. Each of the bones feeding into the spine of the fish represents a specific category of potential contributors to the problem. The team can use pre-determined "cause" categories, such as "people, culture, method, technology", or they brainstorm their own. The team uses the "cause" categories as a prompt for brainstorming potential causes of the problem.

Digital brainstorming and idea evaluation enhances the fishbone technique both in terms of collecting potential causes and by adding the ability to evaluate them in terms of most likely and most serious which provides a prioritization of the most significant issues.

Problem Formulation

The initial step in problem-solving is problem formulation to avoid implementing a solution to the wrong problem and leaving the true problem not remedied. The problem formulation method focuses on identifying, defining, and structuring problems.[2] It involves interpreting the external environment to recognize that a problem exists, considering and selecting one or more perspectives for organizing and defining the type of problem at hand, and proposing an approach to gathering more information or developing an action plan to deal with the problem. It is to stimulate a more thorough consideration of a problem area.

In this approach, the team brainstorms ideas regarding their problem or issue from three perspectives before defining the problem.

1. *Case perspective.* Team members generate and evaluate ideas to describe situations they have encountered that resemble the current situation. This perspective captures the collective experience

of team members. It can also demonstrate the diversity of ways that issues can be linked to other situations. Team members discuss why or how a prior case relates to the current situation.

2. *Category perspective.* Team members brainstorm and evaluate ideas to identify a category, or categories, into which this problem falls. Team members are "classifying" the problem. For example, a lower-than-expected revenue for a product may be a pricing problem (charging more than a competitor), an employee problem (lack of training, motivation), or a production problem (a product gained a poor reputation because of low quality).

3. *Causation perspective.* Team members generate and evaluate their ideas as to what they see as the causes of the problem. They consider the origin of the problem and factors which have led to the current situation.

At this point, the team has considered the current situation from three perspectives, and team members have a sense of how the team as a whole views the problem. They are ready to proceed to define the problem.

Multi-Criteria Decision Analysis

The multi-criteria decision analysis model is useful for teams selecting from a set of alternatives, such as vendors, software packages, or candidates. It is comprised of two steps:

1. Identifying and weighting criteria to evaluate alternatives.
2. Rating each alternative against the weighted criteria to produce an overall score for each alternative.

To develop the model, the team first uses digital brainstorming to create a list of potential criteria to be used to assess the alternatives followed by rating each criterion in terms of its useful value as part of the model. As with other digital evaluation exercises, the averages and ranges provide a starting point for discussion to get clarity about what each criterion means in the context of making the ultimate decision. The average ratings for each criterion can become a relative weight for each one.

With a list of alternatives, an advanced team decision model capability is used to provide team members with the ability to rate each alternative, e.g., Vendor A, Vendor B, Vendor C, against the criteria using a scorecard as illustrated in Figure 3.11. The combined team member evaluations will produce a weighted average score for each alternative.

Criteria	Weight	A	B	C
	X.X			
	X.X			
	X.X			
	X.X			
	X.X			
	X.X			
	X.X			

FIGURE 3.11   Example scorecard for each user to evaluate each of the options (A, B, C) against the weighted criteria.

The combined team member ratings will produce a weighted average score for each alternative which supports the team in making a recommendation, or selection, among the alternatives.

STAGE 4 – CONTINUE OR EXIT?

Upon achieving its purpose, the team reflects on its "work products" and processes. They consider what role they may have going forward. Digital brainstorming and idea evaluation can be used to facilitate conversations regarding "lessons learned", team member reflections, and team member outlook.

DIGITAL TEAMING EXAMPLE: SELECTING A SOFTWARE PACKAGE

This section provides an example of digital teaming for a common business project: selecting a software package, or service, to automate a business process. Before a team is chartered to make a recommendation, the leadership team creates a "vision" to provide the context and purpose of the new software package. It expresses the expected outcomes and high-level changes to the current process. It becomes input to the initiative team's "selection" process using the multi-criteria decision analysis model which is outlined in the following steps:

1. Identify the criteria by which the alternatives will be judged.
2. Weight the criteria by importance.
3. Identify and research alternatives, i.e., CRM packages.
4. Evaluate the top alternatives against weighted criteria.
5. Recommend the best alternative and document the findings.

STEP 1. IDENTIFY CRITERIA

Each team member privately brainstorms ideas for criteria to judge vendor packages. Team member, Dave Brunello, enters "Track sales opportunities" as a potential criterion as shown in Figure 3.12. These criteria will include the types of features the CRM software package should provide to meet the business's needs. They are informed by the "vision" provided by the leadership team.

Each team member rates the list of ideas for criteria on a 5-point scale where "1" equals "strongly disagree" that it should be a criterion and "5" equals "strongly agree" that it should be a criterion as shown in Figure 3.13. These ratings are done by each member anonymously.

Once all team members have completed their ratings, the resulting averages are displayed as shown in Figure 3.14. In addition to the average, the

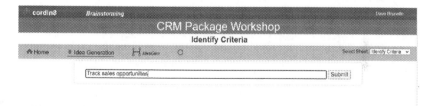

FIGURE 3.12   Each team member enters ideas as potential criteria.

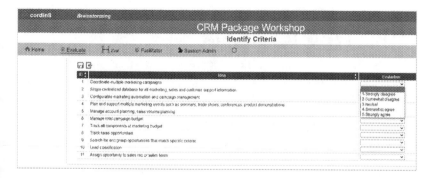

FIGURE 3.13   Each team member rates the list of brainstormed ideas as potential criteria.

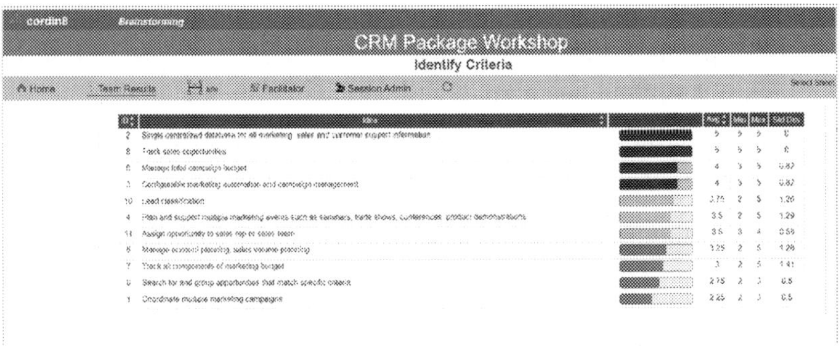

FIGURE 3.14   Average rating results with "min" and "max" range and standard deviation.

minimum and maximum ratings are shown along with the standard deviation. The smaller the standard deviation, the more agreement there is amongst team members regarding the rating. The larger the standard deviation, the greater the disagreement. "Single centralized databases for all marketing, sales and customer support information" and "Track sales opportunities" rated the highest to be included as criteria with consensus among team members.

When there are significant disagreements, such as "Lead classification" where there is a range of "2" to "5", it is often the case that team members do not have the same understanding. Discussion, especially using examples, helps team members gain clarity and arrive at a common understanding. Following these types of discussions, it is useful for team members to conduct another round of rating and review the new results to see if agreement on those items improved. If the criteria list is too large, the team may want to reduce it. Using the sorted list from highest to lowest, those that fall below a threshold, e.g., a rating below "3.5", can be dropped.

STEP 2. WEIGHT CRITERIA

It is often the case that some criteria are more important than others. Weighting them allows their relative importance to be captured in the final calculations. There are different approaches to weighting, but one can utilize the results from the rating exercise by using the average criteria ratings as a guide. In the previous Figure 3.14, the top two criteria have the same score of "5", so they can be assigned a weighting of "20" each. As the next two have an average rating of "4", they can be assigned a lower weighting of "15" each. The next three range from "3.75" to "3.5" and can be weighted "10" each. Those criteria below "3.5" are dropped from the list.

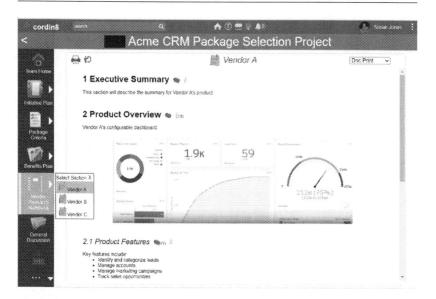

FIGURE 3.15 Vendor Research Notebook for collecting and organizing research on three vendor CRM packages.

STEP 3. IDENTIFY AND RESEARCH VENDORS

Using the list of weighted criteria developed in the prior steps, the team identifies vendor software packages that have capabilities to meet their criteria. Using a collaborative notebook, team members contribute to a shared "notebook" with details of package features, cost, user interface examples, customer feedback, and so forth as shown in Figure 3.15. It is a "shared" knowledge source that team members build together.

Team members can add commentary and replies to share their opinions about each of the packages. While it becomes supporting documentation as part of the team's report for their recommendation, more importantly, it helps to educate team members about the packages under consideration. If there are many vendor packages, the team may want to conduct an exercise to rate the list of vendors to identify the top "three" to invite to provide them with a demo and answer questions.

STEP 4. EVALUATE VENDORS AGAINST CRITERIA

With the research step completed, team members are ready to rate the vendors against the weighted criteria developed in Steps 1 and 2. Each team member

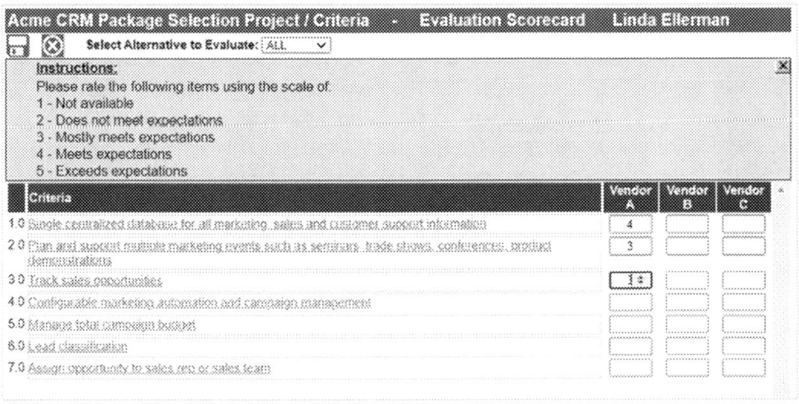

Acme CRM Package Selection Project / Criteria - Evaluation Scorecard Linda Ellerman

Select Alternative to Evaluate: ALL

Instructions:
Please rate the following items using the scale of:
1 - Not available
2 - Does not meet expectations
3 - Mostly meets expectations
4 - Meets expectations
5 - Exceeds expectations

	Criteria	Vendor A	Vendor B	Vendor C
1.0	Single centralized database for all marketing, sales and customer support information	4		
2.0	Plan and support multiple marketing events such as seminars, trade shows, conferences, product demonstrations	3		
3.0	Track sales opportunities	1		
4.0	Configurable marketing automation and campaign management			
5.0	Manage total campaign budget			
6.0	Lead classification			
7.0	Assign opportunity to sales rep or sales team			

FIGURE 3.16 Team member evaluation scorecard.

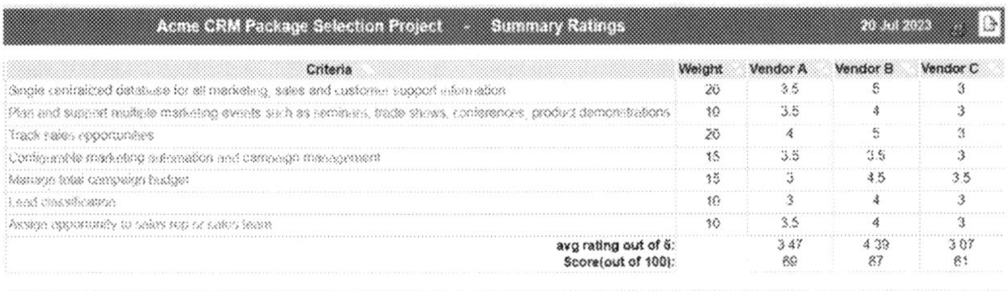

Acme CRM Package Selection Project - Summary Ratings 20 Jul 2023

Criteria	Weight	Vendor A	Vendor B	Vendor C
Single centralized database for all marketing, sales and customer support information	20	3.5	5	3
Plan and support multiple marketing events such as seminars, trade shows, conferences, product demonstrations	10	3.5	4	3
Track sales opportunities	20	4	5	3
Configurable marketing automation and campaign management	15	3.5	3.5	3
Manage total campaign budget	15	3	4.5	3.5
Lead classification	10	3	4	3
Assign opportunity to sales rep or sales team	10	3.5	4	3
avg rating out of 5:		3.47	4.39	3.07
Score(out of 100):		69	87	61

FIGURE 3.17 Weighted average results of vendor ratings against criteria.

gets an "evaluation scorecard" as shown in Figure 3.16. The rating scale used in this example is a 5-point scale, where "1" indicates that the package does not have the capability to a "5" indicating that it exceeded expectations.

STEP 5. RECOMMEND BEST ALTERNATIVE

Once all team members have entered their ratings, the results can be calculated and displayed as shown in Figure 3.17 with Vendor B having the highest score of "87". The results are a weighted average of team member ratings. As done in previous steps involving ratings, the team will discuss the results, explore differences, and re-evaluate before finalizing their recommendation.

The resulting work product from the team is a well-thought-through recommendation that uses an efficient and transparent process. Part of the team's work product will include the output from the team decision support models that provide support for how the team arrived at their recommendation.

SUMMARY

The value of digital teaming lies in its ability to increase participation, visibility, creativity, and diversity of opinion, all of which can increase the team's likelihood of achieving success. With digital teaming, as team members add or delete items, there is a movement toward adopting that information as team-owned. It becomes a team product. Without digital teaming, information may be volunteered, opinions offered, and alternatives evaluated, but the information itself remains member-based with individuals, rather than compiled into a different team product.

NOTES

1 J. P. Guilford, "Three Faces of Intellect", *American Psychologist*, 14(8), 469–479, 1959.
2 The approach described is based on research using the SAMM system at the University of Minnesota.

Scrumfall

4

INTRODUCTION

The authors describe Scrumfall as a team-centric approach for delivering business initiatives, such as launching a new product into the marketplace, implementing a software package in an organization, and integrating company operations following a merger. Scrumfall teams are supported by a Digital Office Complex, as it provides a fast communication structure and visibility for team members, customers, senior leadership, and other stakeholders. Scrumfall draws upon three major methods for delivering projects: Waterfall, Critical Chain, and Scrum.

Waterfall

At the dawn of the computing era in the 1950s, large-scale systems programmers adopted a logical and sequenced method of project management developed during previous decades in the construction industry. It came to be called "Waterfall" because the model of the software development phases was depicted by a sequence of boxes, labeled for each phase, and connected with downward arrows. The resulting diagram resembled a waterfall with the "analysis" phase at the top spilling downward through the subsequent phases of "design", "development", "testing", and "implementation". In the "design" phase, the team creates a hierarchical outline of the system to be developed based on the requirements collected during the "analysis" phase. The first level in the outline is the name of the new system, which is further decomposed into subsequent levels that represent the software modules to be built during the "development" phase. At the lowest level of the outline, the team identifies "activities" to create the modules which are then used to estimate the effort and costs associated with each module as well as create a precedence network that yields project duration.

 DOI: 10.1201/9781003426127-4

Using this approach, substantive changes to the requirements during the later phases must be managed because they can wreak havoc on the schedule and budget, but not incorporating them can risk customer satisfaction.

Critical Chain

To improve project performance, particularly in the construction industry, Dr. Eliyahu Goldratt applied his work on the theory of constraints to project management.[1] He focused on improvements as measured by completion time. He found that the task estimate quality is often very poor. Even with experts, he found that there can be great uncertainty. He identified that 30% of lost time and resources in projects were consumed by the following:

* Multitasking – a resource switching between multiple tasks.
* Student syndrome – waiting until the end to work on the task.
* Parkinson's law – "Work expands to fill the time available for its completion".
* Inbox delays – reading and responding to emails and other non-task work.
* Lack of proper prioritization.

The "Waterfall" method focuses on managing tasks, whereas the critical chain concentrates on managing resources and buffers. In the critical chain approach, buffers are inserted into the schedule to protect the project's due date and avoid the problems associated with lost time described above. The project manager monitors project performance based on the consumption rate of buffers rather than individual task performance to schedule. Unlike the "Waterfall" method, "Critical Chain" does not use task start and stop schedules. Rather, team members involved in a task chain are expected to execute it as quickly as possible.[2] It has been compared to a "relay race". Each task owner in the chain of tasks is expected to perform the task as quickly as possible and without any interruptions. Once completed, the next person in the chain takes the "baton" and performs his assigned task uninterrupted and as quickly as possible and so on until the chain is complete. Users of this approach, especially engineering and construction firms, claim that it has improved project performance by 25% or more.

Scrum

In 1995, Ken Schwaber and Jeff Sutherland presented the *Scrum Software Development Process* at a conference in Austin, TX, as a "better way of team collaboration for solving complex problems".[3] The name of this approach was inspired by rugby, where the team comes together in a Scrum to work together to move the ball forward. In the context of software development, Scrum is where the team comes together to move the product forward. It has since gone on to become an extremely popular approach to software development.

A Scrum team is comprised of three roles: "Product Owner", "Scrum Master", and development team. The "Product Owner" represents the customer's voice in terms of product features. The "Scrum Master" leads and coordinates the development team in the use of the Scrum flow. The development team is comprised of the people with the skills needed to deliver the features identified by the "Product Owner". It is self-directed which means that the members participate in managing the team's work. A critical tool for supporting Scrum teams is a Kanban, which is a Japanese word that roughly translates to "a card that you can see". As illustrated in Figure 4.1, it is a visual display system used to manage and track work as it moves through a process.

In Scrum, small sets of features are developed during periods known as "sprints", which are typically two or four weeks in duration. The "Product Owner" determines the set of features to be developed during a sprint based on the value of the feature to the customer with input from the development team regarding the effort to create them within the sprint timeframe. At the end of the sprint, the newly developed capabilities are demonstrated to the customer for review. These demonstrations often inspire enhancements or new features, which are added to the "product backlog" for consideration in the next or future sprints. Unlike "Waterfall", Scrum welcomes enhancements and new features. They are viewed as bringing additional value to the customer.

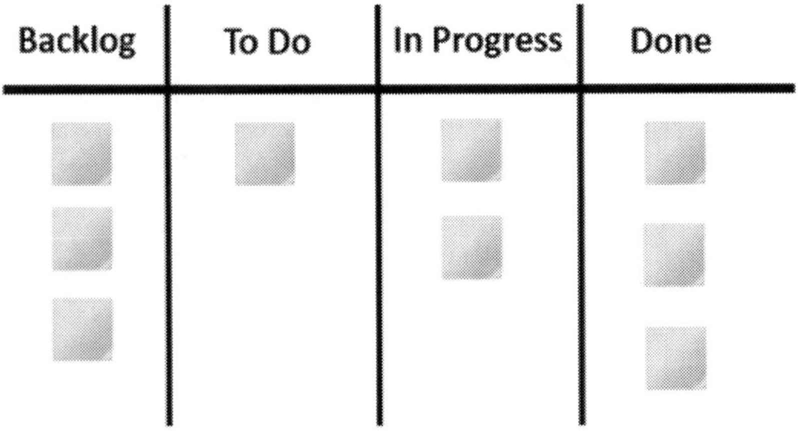

Backlog	To Do	In Progress	Done

FIGURE 4.1   Example of Kanban board.

SCRUMFALL

Scrumfall combines selective qualities of "Waterfall" with the team engagement and customer focus of Scrum. It is also inspired by the learnings from Goldratt's work on the critical chain method to overcome time lost to the "student syndrome", Parkinson's law, and the lack of prioritization. Figure 4.2 is a graphical representation of the Scrumfall workflow, which is described in the subsequent sections. While general descriptions of project management tools, such as scheduling, budgeting, or managing risks are provided, the reader should refer to other sources for an in-depth treatment of them.[4]

A. *Initiative Planning*
   1. Deliverable breakdown structure (DBS) – a breakdown of the deliverables to be produced by the team.
   2. Deliverable package (DP) – the lowest level in any branch of the DBS.
   3. Schedule and budget – estimated timeline and costs of deliverables for the initiative.
B. *Quarterly Planning*
   4. Quarter Objectives (QOs) – at the beginning of a quarter, the team identifies the DPs that they will pursue in the quarter. It results in a list of quarterly objectives.
   5. "Activities" Team Backlog – the list of "activities" needed to complete the DP.

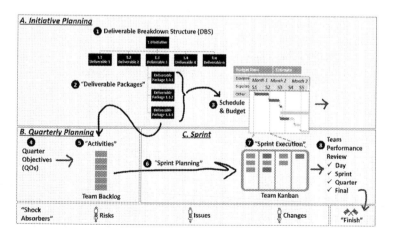

FIGURE 4.2 Scrumfall flow.

C. *Sprint*

6. Sprint planning – at the beginning of a "sprint", a period of two or four weeks, based on their knowledge and experience in producing outputs defined by the deliverable packages, the team "pulls" the "activities" from the Team Backlog that it will target to complete by the end of the sprint. They are placed in the first column, "Not Started", on the team's Kanban.

7. Sprint execution/team Kanban – as team members work on "activities", they move the card representing them across the board from "Not Started" to "In Progress" to "Review" and then "Complete". The goal of the sprint is to move all "activities" in the "Not Started" column to the "Complete" column by the end of the sprint.

8. Team performance reviews – during a sprint, the team's performance is based upon achieving sprint goals and quarterly project objectives. The team reviews its performance with customers, senior leadership, and other stakeholders. They hold daily "stand-ups", sprint reviews, and quarterly reviews.

Shock Absorbers

The team uses the "shock absorbers" to manage things that may arise during their work that will impact their ability to produce the deliverables within the timeframe and budget. These include risks, issues, and changes to the scope. One of the objectives of including leadership, customers, and stakeholders is to keep them abreast of progress and any challenges as well as to get their input on anything that will impact completion.

Finish

At the final review, the team presents a final performance review of the "budget vs. actual" costs and recommendations for future work and any ongoing issues.

DIGITAL OFFICE

Shown in Figure 4.3 is an example of a Digital Office configured to support an initiative team using Scrumfall. It is comprised of eight primary digital office objects:

1. Team Home – share news and upcoming events.
2. Initiative Plan – a collection of collaborative documents to support the planning and management of the initiative, including a performance dashboard, business case, initiative attractiveness, Initiative Charter, stakeholders, DBS, resources, communication plan, risk issues, changes, lessons learned, meetings (log), monthly status report, and final team performance report.
3. Requirements Plan – a collection of collaborative documents to support the management of requirements for the initiative and the "requirements" themselves.
4. Benefits Plan – a collection of collaborative documents to support the planning and management of benefits to be harvested from the initiative.
5. Activities – a Kanban to track initiative activities.
6. Training Notebook – a collection of collaborative documents for developing training materials to be used during implementation.
7. Discussion – an area for "Q and A" that covers broad topics related to the initiative.
8. Approvals – a collection of documents that are managed through an approval workflow, such as an Initiative Charter.

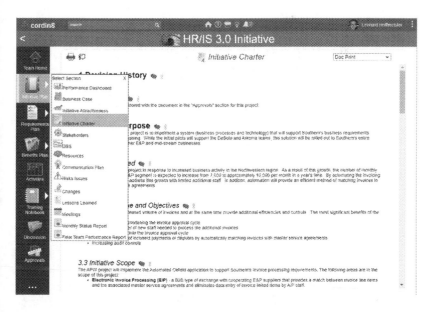

FIGURE 4.3 Digital Office example to support an initiative using Scrumfall.

Having the content and tools to support Scrumfall organized in one place makes it easier for team members to access and make contributions to relevant information quickly. To illustrate, a team member working on the Initiative Charter finds it as part of the Initiative Plan digital office object. Its contents are directly displayed in contrast to the contents being retrieved from a word processing document. The team member may add content, post a question, or attach supporting files to any of the charter's sections. Once it's completed, the collaborative document is saved and routed for approval using the workflow in the "Approvals" digital office object.

When "visiting" this initiative team's Digital Office, it is easy for stakeholders to orient themselves and discover relevant and real-time information for themselves. While periodic summarized reports pushed to stakeholders remain valid, the ability for them to access the team's Digital Office at any time and have the latest content along with the ability to pose questions or provide feedback is a powerful way to engage stakeholders.

The content stored in digital office objects is accessible programmatically. A Digital Office supporting a leadership team that is sponsoring this initiative can retrieve content for real-time reports and dashboards by retrieving data directly from the digital office objects rather than a person extracting data and emailing it to them.

INITIATIVE PLANNING

One of the most effective tools to use for organizing an initiative is the work breakdown structure (WBS), or the authors' preferred term, "deliverable breakdown structure (DBS)". It was initially developed in the 1960s by the US Department of Defense and NASA to plan and manage very large projects; however, it is useful for any size of endeavor. After the initial step of team formation and achieving an understanding of the team's purpose, creating the DBS comes next.

The focus of the DBS is on the "outcomes" of the initiative vs. the "tasks" to achieve them. Examples of planned outcomes include tangible products (e.g., manufacturing plant), intangible products (e.g., software, architectural design, policy), and results (e.g., additional revenue, improved customer satisfaction). Instead of "What are all the 'tasks' we need to do?", the team asks itself, "What are the 'deliverables' we need to produce to achieve our purpose?" Organizing the latter into a hierarchical structure is the process of creating a DBS. Larger, complex deliverables are decomposed, or broken down, into smaller, less complex deliverables. The deliverable that is at the lowest level of a breakdown is referred to as the "deliverable package". Documenting additional information for each deliverable package will enable the team to create a plan to guide their efforts in producing them.

Creating the DBS

Multiple approaches can be used to create the DBS, including bottom-up, top-down, and leveraging previous work. The following is a bottom-up approach using digital teaming capabilities described in Chapter 3. The format for this process is a workshop ideally with members in the same room.

1. Team members brainstorm "deliverables" for the initiative. Deliverables are the "Whats". The team uses nouns and adjectives to document a deliverable. Deliverables do not include verbs. An example of a deliverable is a "Strategic Workshop" and not "Conduct Strategic Workshop". The "activities" will come later in the process.
2. After deleting duplicates that arose during brainstorming, the team clusters similar "deliverables" together and labels them.
3. The team stores the output from "clustering" into an "outline" structure to develop the rest of the DBS. The "cluster name" becomes a level in the DBS hierarchy with the individually named deliverables at the next level down.
4. Depending upon the complexity of the initiative, the team may need to arrange "clusters" together to form a higher-level "grouping", which is also given a name.
5. As the structure begins to take shape, team members are likely to think of additional deliverables that need to be inserted into the structure. It may also lead to re-arranging the structure for better clarity and simplification.
6. The team should include deliverables related to managing the initiative, such as the Initiative Charter, Risk Register, Stakeholder Register, communications plan, and so forth.
7. Once these "groupings" and restructuring have been completed, a top-level item is created and named after the initiative.
8. The team reviews the entire structure to ensure that all deliverables have been identified.

Figure 4.4 is an excerpt of a DBS for implementing a software package. It represents "What" the team needs to produce to fulfill its mission concerning improving HR processes. At this stage, it does not define "how" the team will do it.

```
1.0      HR Software Package Implementation
1.1        Requirements
1.1.1        Process Analysis & Redesign
1.1.2        Requirements Finalization
1.2        Solution Design
1.2.1        Payroll Processing Provider Data Mapping
1.2.2        Data Analysis
1.2.3        Security Analysis
1.2.4        Technical Specifications & Design Document
```

FIGURE 4.4   Excerpt of DBS for implementing HR software package.

The first item is the name of the initiative, "HR Software Package Implementation". The deliverable packages are the items listed at the lowest level of the structure, such as "1.1.1 Process Analysis & Redesign" and "1.2.4 Technical Specifications & Design Document".

Creating the DBS Dictionary

The DBS alone does not provide sufficient information to guide the team in creating the necessary deliverables. Additional information at the deliverable package level enables the team to estimate the relative size and cost of the effort. These additional fields extend the DBS to become a "DBS Dictionary". Determining "When" deliverables are worked on and by "Whom" are the activities that follow the build-out of the DBS Dictionary.

The associated DBS Dictionary fields are as follows:

1. Description – a detailed specification of the deliverable.
2. Scope – defines the boundaries for the deliverable. What is included, and what is not? For example, the scope for a "Strategy Workshop" deliverable might be written as "It will only involve the corporate leadership team". This scope excludes providing a strategy workshop for business units or departments within the company.
3. Sourcing – internal resources or contracted. Will the production of this deliverable be done by internal resources or contracted with an external provider?

4. Stakeholders – individuals, groups, or organizations that have an interest or role to play regarding the deliverable.
5. Assumptions – things considered to be true for planning purposes.
6. Constraints – things that limit options.
7. Risks – anything that has the potential to impact the successful completion of the deliverable. For each risk identified, specify the probability (high, medium, low) of it occurring along with the impact (high, medium, low) should it occur.
8. Communication – identification of communication products, e.g., status reports, to consider when communicating with stakeholders regarding the deliverable.
9. Resources – identification of the skills needed to produce the deliverable.
10. Procurement – identification of equipment, materials, software, services, and so on that will need to be procured to complete the deliverable.
11. Quality – identification of standards necessary for customer acceptance of the deliverable.
12. Activity List/Stories – identification of major activities for those deliverables that will be managed activities, or tasks. For software deliverables using a Scrum approach, it is a list of "stories".
13. Cost – an estimate of the cost for direct labor and procured items needed to produce the deliverable.
14. Confidence level (cost) – high, medium, or low level of confidence in estimate for costs.
15. Time – an estimated time from when work on the deliverable begins to its completion.
16. Confidence level (time) – high, medium, or low level of confidence in estimate for time.
17. Dependencies – the identification of deliverables in the DBS for which this deliverable depends. For example, a deliverable that produces a set of documented interviews needs to be completed before a strategy workshop can be delivered.
18. Additional notes – any additional information that may be useful to anyone who will be working on this deliverable. It might include references to models, articles, and experts.
19. File attachments – files that contain diagrams, illustrations, and detailed notes that will be useful to those who will be working on this deliverable.
20. Collaborative discussion – discussion thread to capture questions, responses, commentary, and interim progress regarding the deliverable.

FIGURE 4.5   Excerpt of a DBS Dictionary form.

A workshop is a useful format to build-out the DBS Dictionary. The workshop agenda can be organized into an hour and 30-minute blocks with breaks following each block. At the start of each block, team members organize into groups of two or three members. If possible, it is recommended to form new groups at each block. Since this piece of teamwork comes early in the team's life, it also serves as a team-building exercise. Changing the members of the groups allows each team member to work with all other team members.

Each group signs up for, or is assigned, a set of deliverable packages to define. Working in parallel, each group thinks through the DBS Dictionary fields described previously for the deliverable packages they have been assigned. At the end of an hour, each group presents their assigned deliverable packages to the whole team who are asked to provide feedback.

Figure 4.5 illustrates an example DBS Dictionary form for the deliverable package, "Process Analysis & Redesign".

Creating Initiative Schedule and Budget

The first step toward creating a schedule is sequencing the deliverable packages in order of precedence. Those that must be completed before another one can start are sequenced ahead. Next, provide a relative size of the effort,

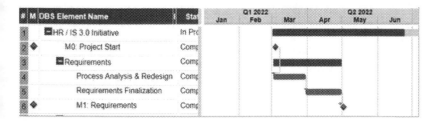

FIGURE 4.6 Example Gantt chart.

such as "Small (S)", "Medium (M)", or "Large (L)". These can later be translated to temporal units, such as "80 hours" for small, "160 hours" for medium, and "480 hours" for large. The purpose is to eliminate debates and discussion time about whether the estimated effort is 70 hours or 85 hours when both are guesses about what it might be. With the precedence network and the translation of the relative sizing efforts to hours, a timeline can be created, such as the one shown in Figure 4.6.

The estimated cost can be calculated based on the type (e.g., financial, engineer, IT), quantity, amount (e.g., 10%, 25%, 100%), and expertise level (e.g., basic, intermediate, advanced) of the resources needed to produce the deliverable package in addition to procurement costs (e.g., consulting services, hardware, software) identified in the DBS Dictionary.

QUARTERLY PLANNING

Using the schedule created in the previous step, the team identifies the deliverable packages to target for completion in the quarter. They become candidates for the "Quarter Objectives (QOs)". Deliverable packages that were not completed by the end of the quarter factor into the objectives for the subsequent quarter. If the schedule indicates that a deliverable package will cross a quarter boundary, the team will need to break it down further to fit within the quarter and undertake the remainder for the subsequent quarter.

As part of "quarterly planning", team members with the appropriate skills and experience identify and sequence the necessary "activities" to produce each deliverable package identified as a quarterly objective. They start with the "Activity List" developed during the build-out of the DBS Dictionary. They may add, modify, or delete them based on their expertise and feedback received from customers during the review of the QOs. This list of activities is captured in the Team "Backlog".

FIGURE 4.7   Scrumfall Kanban.

SPRINT

The "sprint" is a period of elapsed time of two or four weeks. It starts with sprint planning followed by sprint execution. It is the period the team actively works to produce the outputs as defined by the deliverable package. The team uses the Kanban, as shown in Figure 4.7, to manage each sprint.

Sprint Planning

"Sprint planning" takes place at the beginning of the sprint. Team members identify the "activities" in the "Backlog" that can be completed by the end of the sprint and "pull" them from the "Backlog" column across to the "Not Started" column on the Kanban. They commit to completing them within the sprint period.

Sprint Execution

When sprint execution begins, team members sign up for "activities" that are in the "Not Started" column, move them across to the "In Progress" column, and begin working on them. When they have finished with the "activity", they move it to the "Review" column for quality review and acceptance. When an "activity" is blocked, which means that team members are waiting on responses or actions from others, they work on another "activity" underway or move another "activity" from "Not Started" to "In Progress". At any time, team members, customers, and other stakeholders can view the state of the team's work by viewing the Kanban. They can see how many "activities" are currently being worked on, how many are waiting to start, and how many are waiting to be reviewed.

Team Performance Review

There are three team performance review frequencies targeted to different audiences: daily, sprint, and quarterly.

Daily Review

At the beginning of each day during the sprint, the team conducts a daily review, which is a brief meeting that is no longer than 15 minutes. Team members provide an update on the progress of their "activities" and any issues that are, or are foreseen to be, preventing successful completion. These brief updates are captured as part of the "activity" item on the Kanban. It enriches the documentation for the work, allows customers and stakeholders to stay abreast of the progress of the specific activities, and may be referenced at a future date.

One of the tools used to review team performance is the sprint burn-down chart which is updated daily. It is a two-line chart that compares a forecast of the number of "activities" remaining to be completed each day to the actual number that has been completed. The "forecast" starts on day zero with the total number of "activities" in the sprint. It "burns down" each day using the formula of total activities in the sprint divided by the number of days in the sprint. Figure 4.8 illustrates a sprint with 20 "activities" in a 20-day period.

The "Forecast" line starts at "20" on day zero and is reduced to "19" by day one, "18" by day two, and "0" on day 20. The "Actual" line also starts with the total number of "activities" in the sprint on day zero. Each subsequent day, the total number of "activities" remaining is calculated by subtracting the number of "activities" that moved to the "Review" column that day from the "Actual" remaining count of the previous day. Continuing with the example, the "Actual" line starts at "20" on day zero. Although team members were working on "activities" on the first two days, none of them reached the "Review" stage. On day three, two "activities" moved to "Review", which is reflected in the "Actual" line which charts "18" on day three. When the "Actual" line is above the "Forecast", the team is considered behind schedule. If the team is significantly behind as they move past the halfway point of the sprint, they need to understand why they are, explore alternatives to get back on track, and determine their approach for completing the sprint.

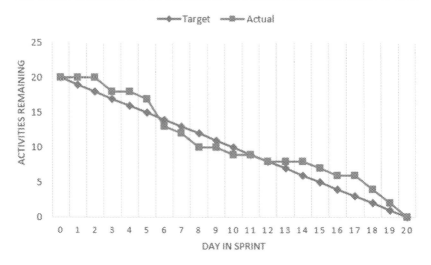

FIGURE 4.8   Sprint function burn-down chart example.

Sprint Review

At the end of a sprint, the team reviews its work products and performance with customers and stakeholders. The team presents work products created from "activities" that have moved to the "Review" column. As customers approve them, the "activity" is moved to the last column, "Complete". The goal for a sprint is to move all "activities" from the first column of the Kanban to the last column.

The team presents the sprint "burn-down" chart, schedule, and budget in reviewing its performance during the sprint. They discuss any issues that arose during the sprint, especially those that continue to be a barrier to completing "activities" in a timely fashion. It is an opportunity for customers and stakeholders to provide feedback to the team regarding the quality of the work and the team's performance to date. This feedback helps the team in planning the next sprint.

This review is powerful because it encourages customers to participate in the success of the initiative. They provide feedback on the targeted deliverable packages which will aid with informing team members working on them in shaping their specific work plans. Knowing the timeline and budget, they can make trade-offs between deliverable packages in terms of scope and quality.

Quarterly Review

The team meets with the leadership team to review the Quarter Objectives and the status of the schedule, budget, and issues. Figure 4.9 is an example report to show the status of the schedule and budget for the initiative. The "S-curve

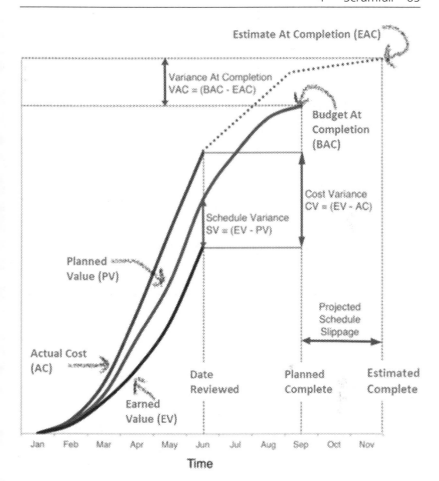

FIGURE 4.9   Example S-curve report.

report" indicates whether the initiative is ahead or behind schedule by comparing "Earned Value (EV)" to "Planned Value (PV)" shown on the chart. It also indicates whether the total initiative budget will be over- or underspent by comparing "Budget at Completion (BAC)" to "Estimate at Completion (EAC)". These calculations utilize the estimated costs for "deliverable packages" found in the DBS Dictionary along with the "deliverable package" status updated during execution and team member time recorded against the "deliverable packages".[5]

The team presents the status of their Quarterly Objectives to customers. This step is powerful by encouraging customers to participate in the success of the initiative. Knowing the timeline and budget, they can make trade-offs between deliverable packages in terms of scope and quality.

SHOCK ABSORBERS

Throughout the initiative, team members will need to respond to unplanned events that will impact their work, such as a risk occurrence, an issue, or a request to change the scope of their work. As part of building-out the DBS during Initiative Planning, the team identified and evaluated risks. For those risks that were evaluated as having a significant to catastrophic impact and are also highly likely to occur, the team will prepare mitigation and response plans which must be included in sprints and the various team performance reviews along with any issues that arise. Changes that will impact the team's scope must be analyzed for their impact and approved by leadership. One of the primary objectives of engaging customers, senior leadership, and other stakeholders throughout the initiative is to get their assistance and support when issues or changes arise that will impact the cost, time, or quality of deliverables.

FINISH

The final team performance review is for team members to engage with customers, senior leadership, and other stakeholders regarding the initiative in totality. The team presents the final deliverables, budget vs. actual, lessons learned, and recommended actions, if any, for future work and outstanding issues.

SUMMARY

Managing complex initiatives, such as drug development, facility construction, or scientific research, is fundamentally about coordinating people with the appropriate knowledge, skill, and experience to create and assemble the identified deliverables to produce an outcome, result, or product. Scrumfall is a team-centric approach to business initiatives. It takes lessons from Waterfall, Critical Chain, and Scrum. It leverages a digital office complex as described in Chapter 2 to provide visibility into the team's work by team members, customers, senior leadership, and other stakeholder groups.

NOTES

1 Eliyahu M. Goldratt, *Critical Chain*, The North River Press, 1997.
2 For additional information, see Gary L. Richardson and Brad M. Jackson, *Optimizing Project Work, Management, and Delivery*, CRC Press, 2023.
3 See Scrum.org at www.scrum.org/learning-series/what-is-scrum.
4 See Gary L. Richardson and Brad M. Jackson, *Project Management: Theory and Practice*, CRC Press, 2017, for more in-depth coverages of these topics.
5 See Gary L. Richardson and Brad M. Jackson, *Optimizing Project, Work, Management, and Delivery*, CRC Press, 2023, for more in-depth coverage.

Digital Governing 5

INTRODUCTION

"Digital governing" builds on digital teaming skills described in Chapter 3 to enhance strategic thinking, change management, and oversight using the capabilities of the Digital Office Complex. Figure 5.1 is an artistic representation of "digital governing". At the center is an arrow representing the leadership team looking from their current position toward their "vision" located at the top of the diagram. At the arrow's base are "strategic thinking" frameworks, including Scenario Planning, SWOT, SOAR, and PESTLE analysis. These frameworks are used to create content that becomes input for developing "Strategy Release 1.0" and subsequent strategy releases. Building on these are "policies" to govern the behavior of the organization, "operations" to inform leadership regarding its business performance for sales revenues and operating expenses, and "KPIs" to provide the progress and status of their strategy toward achieving their vision. Moving out from these are starting from the left:

- Initiatives – teams delivering outcomes that are aligned with the strategy and aiming to move the organization from its current position toward the "vision" as depicted by the "arrow" in the center.
- Stakeholder assumptions – stakeholder expectations critical to the organization's success or failure.
- Enterprise risks – risks, or events, across the enterprise that have the possibility of having an adverse impact on achieving the organization's vision.
- Unknown "unknowns" – information or events that are not known nor can be anticipated to be known. "We do not know what we do not know".
- "Force Field Analysis" – change frameworks, such as "Force Field Analysis", to assist leaders in focusing on the aspects of the strategy that will have significant impacts by analyzing drivers that will provide momentum toward the change objective as well as those that will be headwinds against it.

DOI: 10.1201/9781003426127-5

- Competitors – intelligence about competitors and the competitive environment.
- Customers – knowledge about customers, their needs, and satisfaction with the organization's products and services.
- Innovation/Discovery Zones – teams exploring, prototyping, and learning about needs and capabilities for the future.
- Operations teams – teams delivering operational support and continuous improvement for the organization's "big ideas".

The "Digital Office" for a leadership team supports the creation, evolution, and storage of this content. The "Complex" enables the leadership team to have access to content from other teams, such as the "Initiative" teams, "Operations" teams", and "Innovation/Discovery Zone" teams. All this information establishes the organizational context from which the leadership team members can draw upon to guide the organization.

STRATEGIC THINKING AND CHANGE

SWOT analysis, stakeholder analysis, Force Field Analysis, and Scenario Planning are among the techniques used in strategic planning and change

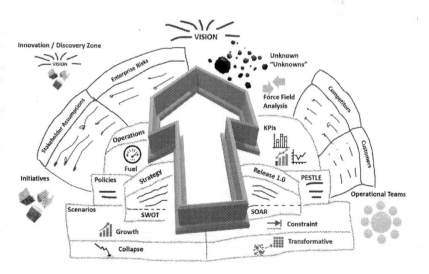

FIGURE 5.1 Artistic interpretation of "digital governing".

management. They are typically organized as workshops with members physically in the same room led by a facilitator and supported by paper-based tools, such as "flip charts" and "sticky notes". Digital brainstorming and team decision support capabilities can improve these workshops through increased diversity of opinion, creativity, and greater participation of the members. In addition, it enables the participation of team members who cannot physically be in the same room where the use of paper-based tools is not practical.

The following sections describe how the capabilities of the "Complex" can be used to improve the quality of outcomes for these techniques.

SWOT Analysis

To conduct a SWOT analysis, the team, or facilitator, will set up four digital "flip charts" one each for "strengths", "weaknesses", "opportunities", and "threats". The team starts by having members brainstorm ideas for the "strengths" of the organization, or business. As has been described, this digital idea generation phase is anonymous. It is also done privately such that team members do not see the combined list of ideas until everyone has finished brainstorming. After allowing for a brief period, such as 30 seconds to a few minutes, the combined list of ideas is displayed on a screen for all team members to review. Once the combined list is shown, there will almost always be duplicate ideas, which need to be deleted. There will also be similar ideas that the team will need to discuss to determine if they are the same or if there are enough subtleties that make them different. Once they have a list of unique ideas, the team is ready to prioritize the list using one of several different idea evaluation techniques. This process is repeated for "weaknesses", "opportunities", and "threats".

For illustration purposes, there is a fictional university located in an urban area that recently built an instructional site in one of the city's suburban areas. As a step in the process of developing a strategic plan for the new entity, the "Steering Committee" conducts a SWOT analysis. Having completed the activity of brainstorming and prioritizing the strengths of the new entity, the Committee moves to brainstorm "weaknesses".

Figure 5.2 shows the unique list of ideas of weaknesses generated by Committee members with a field next to each idea for evaluation purposes. The Committee chose to use the "points allocation" method. In this case, each member is provided with "10" points to allocate across the list of ideas for "weaknesses". Team members may allocate all "10" to a single "weakness" or distribute them in any manner across multiple ones, but the points totaled for all ideas cannot exceed "10".

Once Committee members have completed their evaluations, the results are displayed which sum the points for each idea as shown in Figure 5.3. These

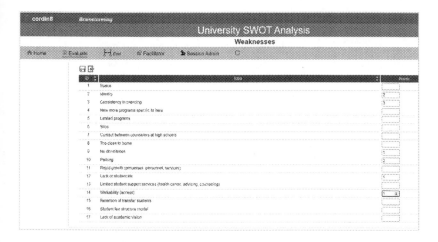

FIGURE 5.2 Team member screen for allocating points to "weaknesses".

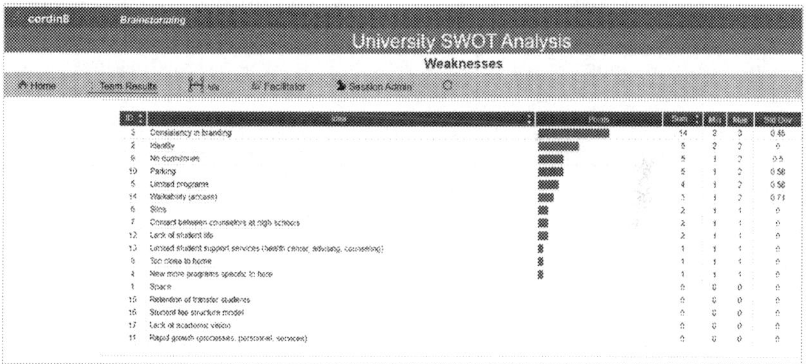

FIGURE 5.3 Results from team member evaluation of "weaknesses" in SWOT analysis.

results show that the Committee members agree that "Consistency in branding" stands out as the most significant weakness with "14" points followed by a related idea, "Identity" with "8" points. Similarly, members were in complete agreement that the bottom five weaknesses displayed were not significant, as they garnered no points. In the few minutes that it took for the members to evaluate the ideas and review the results, they had a prioritized list of "weaknesses".

To get a broader perspective, the Committee sponsored three additional SWOT workshops for members of the community, staff, and faculty. To make it convenient for participants, these workshops were conducted remotely so participants did not have to incur travel time to the location. Digital brain-storming made it possible. They generated and evaluated ideas remotely. When the results were shown, they were displayed on each participant's screen to use in focusing the discussion. With prioritized lists from each of the additional groups, the Committee looked at each to see areas of agreement across all groups as well as identify "weaknesses" they had not considered. Taken together, they now have a solid SWOT analysis.

Stakeholder Analysis

Stakeholders are constituents who have an interest or concern in an organization, and each one has certain expectations of the organization. When the organization does a reasonable job of satisfying a particular stakeholder's expectations, the organization can expect to acquire support and resources from that stakeholder. Effective organizations are successful in acquiring resources and support on a sustained basis by managing the expectations of their stakeholders.

Organizations have many stakeholders, and each stakeholder will likely have many expectations. Because the organization has limited resources, the leadership team must identify a few stakeholder expectations critical to the organization's success or failure.

The process of stakeholder analysis is designed to assist the leadership team with identifying crucial assumptions about stakeholder expectations so that they can create strategies to ensure the organization's success. Digital brainstorming and team decision-support capabilities can support the following three-step approach to stakeholder analysis using a similar method as described previously for SWOT analysis.

1. *Identify stakeholders.* Team members brainstorm the stakeholders of the organization. Teams typically finalize their list through a process of discussion and revision of the initial brainstormed list.
2. *Identify assumptions about each stakeholder.* Stakeholder assumptions are the beliefs or opinions about what stakeholders expect from the organization. An example is that students expect the university to provide good job placement. A digital "flip chart" is created for each stakeholder, so team members can brainstorm potential assumptions for each one.
3. *Evaluate stakeholder assumptions.* The team will likely generate many assumptions about stakeholder expectations. Since it is not feasible to address all of them, it is necessary to identify the most important ones. Two different criteria have been found useful in the evaluation of stakeholder assumptions. The first is "importance to

stakeholder". How strongly does the stakeholder feel about what is implied by the assumption? For example, the previously mentioned example regarding students and the university. Evaluating this assumption on importance to stakeholders is the equivalent of asking the following: Do the students really expect the university to provide job placement? The second is "importance to strategy". This is the equivalent of asking the following questions: Is what is implied by the assumption very important for the organization's strategy? Should the university worry about providing job placement for students?

Having team members rate each assumption on a 10-point scale on both dimensions will produce an average rating for each assumption and plotted as illustrated by Figure 5.4. Those that appear in the upper right quadrant are assumptions

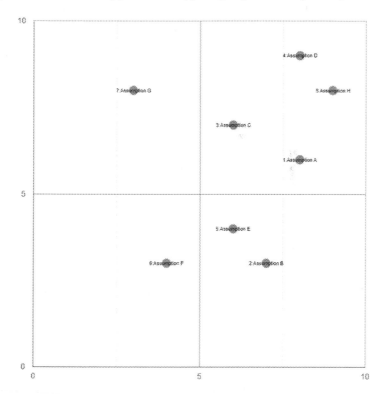

FIGURE 5.4   Stakeholder assumptions matrix.

"D", "H", "C", and "A" that scored high on both dimensions and are seen by the team as crucial to the future success of the organization. Assumptions that score high on one dimension but low on the other are not as important and may not warrant further discussion. Because assumption "F" scored low in both dimensions, it appears in the lower left-hand quadrant and does not merit further discussion by the team. As with all team evaluation results, the team should review the extent to which members agree with the average result.

Force Field Analysis

Force Field Analysis, which was developed by sociologist Kurt Lewin in 1951, is a tool that organizations may utilize in determining the importance, impact, and influence of various factors before implementing a significant change to the organization.[1] This method helps leadership teams assess the factors that are both driving a specific change and those opposing it. Based on this analysis, they can explore actions that may strengthen the "driving forces" and weaken the "restraining forces". The following steps utilize digital brainstorming and team decision-support capabilities to support this analysis.

1. *Identify the change objective.* The team starts the process with the change objective.
2. *Brainstorm the forces that are driving change ("driving forces").* Using digital brainstorming for team members to identify potential forces that are compelling change. If there are many ideas generated, the team should work to reduce the list to the most important ones. The "points allocation" technique for idea evaluation is an efficient means to do so.
3. *Evaluate the "driving forces".* Team members rate each "driving force" on a 5-point scale. Once the results are displayed, team members discuss the results to determine the ratings.
4. *Brainstorm the forces that are opposing change ("restraining forces").* The team uses the same process as described in Step 2, but brainstorm forces that are opposing the change.
5. *Evaluate the "restraining forces".* The team uses the same process as described in Step 3.

Figure 5.5 is an example of a "Force Field" analysis. It places the change objective, "Upgrade factory with new manufacturing equipment", in the middle with the "driving forces" on the left using arrows pointing toward the change objective and the "restraining forces" on the right using arrows pointing toward the change objective. The ratings, which are averages, for each force represent the strength of the force.

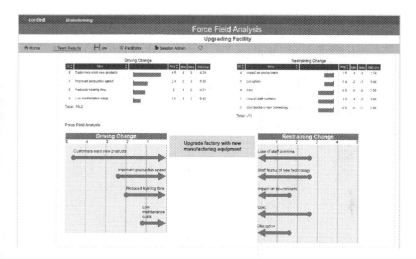

FIGURE 5.5   Force Field Analysis chart.

Overall, the "driving change" forces garnered an average total of "10.2", while the "restraining change" forces received an average of "-11", indicating that the team believes that greater forces are standing in the way of the change. The chart can then be used as a prompt to brainstorm and prioritize actions to strengthen the "driving forces" and weaken the "restraining forces".

Scenario Planning

"With over fifty foreign cars already on sale here, the Japanese auto industry isn't likely to carve out a big slice of the US market for itself" (*Businessweek*, August 2, 1968).

"I think there is a world market for about five computers" (Thomas J. Watson, chairman, IBM, 1943).

"There is no reason for any individual to have a computer in their home" (Ken Olson, president, Digital Equipment Corporation, 1977).

Paul J. H. Schoemaker observed "how frequently smart people have made the wrong assumptions about the future with great certainty".[2] He goes on to describe the value of Scenario Planning as how companies can combat overconfidence and tunnel vision common to so much decision-making. It can be used to stimulate thinking among leadership teams to consider changes that they likely would have ignored. With this process, they identify basic trends and uncertainties that are then used to construct a variety of future scenarios in a narrative form.

When developing scenarios, it is useful to invite people who are external to the organization to participate, such as major customers, key suppliers, consultants, and academics. Incorporating their perspective and ideas helps the leadership team see the future more broadly in terms of fundamental trends and uncertainties.

The following is an example process for Scenario Planning using Schoemaker's ten-step approach and supported by digital brainstorming and team decision-support capabilities plus a team "notebook". It is organized into four workshops followed by an asynchronous workshop, where team members can participate at their convenience rather than a pre-determined date and time.

Workshop #1

- *Step 1 – Define Scope.* The team determines the timeframe and scope of analysis in terms of products, markets, geographic areas, and technologies. Using digital brainstorming, each team member generates ideas about what knowledge would be of the most interest to the organization. Schoemaker suggests that team members consider the past and think about what they wish they had known then and what they know now. What have been past sources of uncertainty and volatility? Look back over the past ten years at the changes that have occurred in the department, enterprise, industry, region, country, and even the world. The leadership team should anticipate a similar amount of change or even more in the next ten years.
- *Step 2 – Identify Major Stakeholders.* Who will have an interest in these issues? Who will be affected by them? Who could influence them? Using digital brainstorming, team members generate stakeholders and then rate them on a scale from "1" to "10" on two dimensions: power and influence. It will result in the creation of a 2 × 2 matrix with those stakeholders rated the highest in power and influence appearing in the upper right quadrant, who become the focus for the remaining exercises. Schoemaker suggests conducting this exercise in two rounds. The first is an evaluation of how each stakeholder rated in terms of these dimensions ten years ago, and the second is how they rate today. The team can then compare and discuss the changes in position and why.

Workshop #2

- *Step 3 – Identify Basic Trends.* What political, economic, societal, technological, legal, and industry trends are sure to affect the issues the team identified in Step 1? The team uses digital brainstorming to create a prioritized list for each of these categories in the same fashion as the SWOT Analysis described previously.

Workshop #3

- *Step 4 – Identify Key Uncertainties.* What events, whose outcomes are uncertain, will significantly affect the issues the team is concerned with? Team members brainstorm uncertainties followed by rating them to identify the top ones.

Workshop #4

- Step 5 – Construct Initial Scenario Themes. Once the team has identified trends and uncertainties, they have the main ingredients for scenario construction. Two approaches for consideration are brainstorming "starter" scenarios followed by grouping similar ones to create and evaluate themes. This is the "snow card" technique described in Chapter 3. Another approach is to start with a set of alternative future archetypes, such as the classic ones created by Jim Dato at the University of Hawaii:[3]

  1. Growth – a future in which current trends and conditions, both good and bad, continue to grow as they have in the past.
  2. Constraint – a future in which scarce resources force societies to contend with limitations.
  3. Collapse – a future in which systems are strained beyond the breaking point, causing system collapse and social disarray.
  4. Transformative – a future in which fundamental change in technology and values signals a break from the past.

After brainstorming the scenarios, the team saves them to the team notebook in their Digital Office to build them out.

Asynchronous Workshop

- *Step 6 – Check for Consistency and Plausibility.* This step is a review and edit of the initial scenarios stored in the team notebook.

- *Step 7 – Develop Learning Scenarios.* The initial scenarios provide future boundaries, but they may be implausible, inconsistent, or irrelevant. The goal is to identify strategically relevant themes and then organize the possible outcomes and trends around them.
- *Step 8 – Identify Research Needs.* There may be a need to do further research to flesh out the understanding of uncertainties and trends.
- *Step 9 – Develop Quantitative Models.* After completing additional research, the team should reexamine the internal consistencies of the scenarios and assess whether certain interactions should be formalized via a quantitative model.
- *Step 10 – Evolve toward Decision Scenarios.* Finally, in an iterative process, the team must converge toward scenarios to use to test their strategies and generate new ideas.

Using their team "notebook", or "Scenario Notebook", as illustrated in Figure 5.6, each scenario can be built out using Steps 6–10 without meeting at the same time nor in the same place, referred to above as an "asynchronous workshop", though it is productive for the team to have short meetings to synchronize the overall effort. The outline shown is the scenario narrative supported by sub-sections for strategic challenges, core capabilities, and quantitative models.

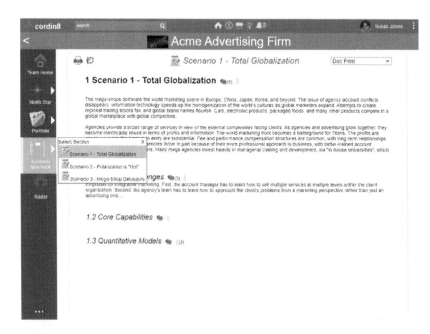

FIGURE 5.6   Scenario Notebook for team members to build-out the scenarios.

All team members, as well as invited participants, can read and add commentary and pose and respond to questions. To pose or review comments, the team member clicks on the "discussion bubble" next to the title of the section or sub-section. In this example, "1 Scenario – Total Globalization" and "1.1 Strategic Challenges" have "1" and "3" comments respectively. In addition, a team member may add supporting documentation to any of these sections by attaching files through the "paperclip" icon. "1.3 Quantitative Models" has a "2" next to the "paperclip" icon, indicating that there are two files associated with the sub-section.

Once the team is satisfied with the scenarios, they can extract the narratives to publish and disseminate. Retaining the "thinking" work that went into the development of the scenarios enables current or future team members as well as others that become involved in Scenario Planning to review the conversations that led to the creation of them.

STRATEGIC OBJECTIVES

There are different frameworks for managing strategic performance. The use of digital brainstorming and team decision support to create strategic objectives for any of them is similar. The balance scorecard framework is one such example. It typically uses four dimensions: "Financials", "Customers", "Internal Processes", and "Learning and Growth". Because digital brainstorming is so efficient, it is an opportunity to extend participation in the creation of strategic objectives to the next level of managers beyond the leadership team. They have more in-depth knowledge of their specific area of responsibility that can assist in shaping the objectives. It also contributes to increased clarity of the strategy as well as commitment to it.

For each dimension, all participants generate ideas for potential strategic objectives and then rate them as to their contribution to moving the organization toward their vision. The top-rated ones become the objectives.

MANAGING BY WALKING AROUND DIGITALLY

HP's Bill Hewett and David Packard created the concept of "managing by walking around" (MBWA) later popularized by authors Tom Peters and Robert Waterman.[4] It involves managers wandering around the workplace without a

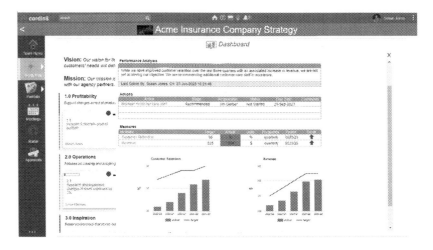

FIGURE 5.7   Leadership Digital Office Complex dashboard.

pre-defined agenda. The purpose is to listen and learn from employees in their work environment. Members of the leadership team can view the real-time progress of their strategy in their Digital Office as shown in Figure 5.7. Any member can navigate via the Organizational Directory, as described in Chapter 2, to one or more Digital Offices to "walk around digitally" to learn about the team's work and the challenges they might be facing. Because the Digital Offices within the Complex have a common interface, "visitors" can easily orient themselves to know where to find certain types of information quickly instead of wasting time having to figure out how each team organizes its work. Scanning the conversations related to issues, for example, may prompt a leadership team member to connect an expert from another part of the organization to the team to assist in addressing an issue.

> Hunter, Acme CEO, was on a call with a board member, Lauren. Knowing she would want to stop by for a briefing on the company's large digital transformation initiative, Hunter navigated on his tablet to the initiative team's Digital Office to take a glance at their dashboard, which consolidated data and content from the various teams working on it. They had made good progress since she had last visited with the Initiative Manager two weeks ago. He relayed to Lauren, "The team had just completed one of the larger deliverables, but I would like to get your thoughts regarding a major risk". After he finished his call, he messaged

Rachel, the Initiative Manager, who was traveling and in a time zone that was six hours ahead. He asked her to get back to him with updates on the major risk and status regarding other issues. Thomas replied, "I'll check with all the leads and get back to you". Meanwhile, Hunter drilled deeper into the Digital Offices of all the teams involved, which made him more informed. It would help him ask better questions of Rachel when they talked. He would be well-prepared for the meeting with Lauren, which was in less than an hour.

SUMMARY

Digital governing facilitates "strategic thinking" so leadership teams can provide guidance and direction for teams in the organization. It supports the leadership team in assessing performance and decisions regarding course corrections. Through capabilities that facilitate team members located in different geographies and major stakeholders from different organizations, they increase the diversity of opinions in the "thinking" process that shapes strategy and informs the selection of initiatives. Digital governing enables the ability to view any team within the digital office complex or a consolidated view of the teams contributing to an enterprise objective which is far superior to individual teams having to extract, format, and email reports.

NOTES

1 Kurt Lewin, *Field Theory in Social Science*, Harper and Row, 1951.
2 Paul J. H. Schoemaker, "Scenario Planning: A Tool for Strategic Thinking", *MIT Sloan Management Review*, January 15, 1995.
3 Jim Dato, "Alternative Futures at the Manoa School", *Journal of Futures Studies*, 14(2), 1–18, November 2009.
4 Tom Peters and Robert Waterman, *In Search of Excellence: Lessons from America's Best-Run Companies*, Harper & Row, 1982.

Artificial Intelligence

6

INTRODUCTION

Using a forklift to move a pallet of bricks, a jack to lift a car to change a flat tire, or a computer to produce invoices for ten million customers monthly are all examples of machines that make tasks possible that otherwise would be impossible. Another lever is artificial intelligence (AI), which is an area of computer science, inspired by simulating the human brain. It uses mathematics, statistics, and calculus to process data, especially enormous volumes of data, to predict outcomes very quickly.

The term "artificial intelligence" was first coined by John McCarthy in 1956 when he held the first academic conference on the subject. In the quest to explore how to design algorithms and data structures to simulate the brain, specialties, such as machine learning, deep learning, natural language processing, and generative AI have emerged as tools to accomplish what would otherwise be the impossible task of processing enormous volumes of data quickly to predict outcomes. It has proven successful in the areas of robotics, fraud detection, predicting customer behavior, facial recognition, and autonomous vehicles to name a few. The following sections will provide an overview of key AI areas to augment teaming.

MACHINE LEARNING

In 1952, Arthur Samuel, an IBM computer scientist, coined the phrase "machine learning" to capture the notion of a program that could learn and improve as illustrated by his checkers-playing program which ultimately beat a checkers master in 1962.[1] Machine learning is a subfield of artificial intelligence that uses various self-learning algorithms that derive knowledge from

DOI: 10.1201/9781003426127-6

data to predict outcomes. The algorithms also adapt in response to new data and experiences to improve their efficacy over time.

Neural Networks and Deep Learning

Neural networks mimic how neurons in the brain signal one another. They are a subset of machine learning and are the backbone of deep learning algorithms that enable the use of large data sets. They were first proposed in 1944 by Warren McCullough and Walter Pitts, who were researchers at the University of Chicago.[2]

Neural networks are made up of node layers – an input layer, one or more hidden layers, and an output layer as shown in Figure 6.1. Each node represents an artificial neuron that connects to another with each connection having an associated weight and threshold value. When one node's output is above the threshold value, that node is triggered ("fires") to send its data to the network's next layer. If it's below the threshold, no data is sent. If the neural network has more than three layers, it is considered a deep neural network (DNN), and deep learning refers to the depth of the layers in the neural network.

Neural network training is the process of teaching a neural network to perform a task. The first trainable neural network, the "perceptron", was

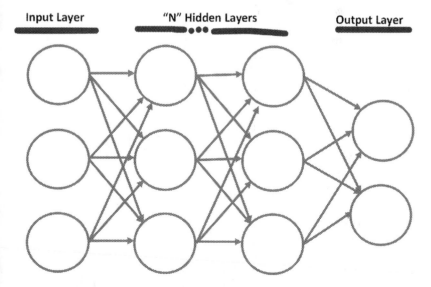

FIGURE 6.1   Neural network example.

demonstrated by Frank Rosenblatt, a Cornell University psychologist, in 1957.[3] It had only one layer with adjustable weights and thresholds, sandwiched between input and output layers. Neural networks learn by initially processing several large sets of labeled or unlabeled data. By using these examples, they can then process unknown inputs more accurately. When a neural network is being trained, all its weights and thresholds are initially set to random values. The training data is provided to the input layer and passes through the succeeding layers, multiplied and added together, until it finally arrives, radically transformed, at the output layer. During training, the weights and thresholds are continually adjusted until training data with the same labels consistently yield similar outputs. Training helps improve their accuracy over time. Once the learning algorithms are fined-tuned, they enable a user to very quickly classify and cluster data. Google's search algorithm is an example of a neural network. Advances in neural networks are the result of computer hardware for video gaming that requires graphics processing units (GPUs), which use thousands of simple processing cores on a single chip, to support its complex imagery.

Generative AI

Generative AI describes algorithms that can be used to create new content, including audio, code, images, text, simulations, and videos. In contrast to the early chatbots of the 1960s, such as ELIZA, that used pre-defined rules to provide responses to a user, contemporary generative AI models do not. These neural networks are trained on vast amounts of data, which are used to develop a representative model that, in turn, is used to generate novel content in response to prompts. OpenAI's ChatGPT can generate text and DALL-E can generate images.

AI Hallucinations

An AI hallucination is when an AI model generates incorrect information but presents it as if it were fact. In early 2023, Google illustrated its new generative AI capability called Bard. It included the question, "What new discoveries from the James Webb Space Telescope can I tell my 9-year-old about?" Among its response of three bullet points was "took the very first pictures of a planet outside of our own solar system". However, according to NASA's website, the first picture was imaged in 2004 by the Very Large Telescope (VLT), operated by the European Southern Observatory in the Atacama Desert of northern Chile.

AI hallucinations can have business performance impacts. For example, an AI model might incorrectly identify major issues with an initiative, leading to unnecessary, even costly, management interventions.

AI-ENABLED DIGITAL BRAINSTORMING AND TEAM DECISION SUPPORT

The following sections describe capabilities aimed at augmenting teaming. It could be used to significantly reduce administrative overhead associated with digital brainstorming and team decision support.

Presenting Similar Ideas for Review

Because team members brainstorm privately first, it is common to have ideas that are similar in meaning when all team members' ideas are combined and displayed. Before moving to idea evaluation, the team must first cull the list to only include those that are unique. An example of two differently worded ideas that have similar meanings is "Communicate with the departments" and "Meet with the business units". "Communicate" is a broader term, which can include the act of "meeting" in addition to other acts, such as "emailing" and "calling". On the other hand, "meet" is a very specific action. AI can analyze the team's combined list of ideas and present those that have similar meanings. Following a discussion, the team can select one or the other, create a new idea that combines the intent of both ideas, or rewrite each one to be clearer and more distinct.

Clustering Ideas

After a team brainstorms ideas, the list can have 50 to a hundred ideas or more. Even after deleting exact duplicates, the list may still be very long. AI can be used to cluster similar ideas along with a recommended label for each grouping. For each cluster, the team can review the list of ideas to determine if they agree with the grouping and label. As they move through all the clusters, they may move ideas from one to another as they see fit, rename a cluster, or reorganize it. AI makes the initial "clustering" activity very efficient. Additionally, it is very difficult to conduct this activity when team members are remote without this form of assistance. A powerful example of this capability is applying to a large list of ideas for features of a new product.

Feedback on Team Idea Evaluation Results

After each team member privately rates the list of ideas, AI can be used to provide feedback to the team regarding the results, such as a prioritized list to guide the team in a discussion emphasizing where members disagree. It could suggest explanations for why the differences of opinion might exist that the team can consider during their discussion, which removes a barrier to identifying potential explanations of controversial results.

Multi-team Analysis

In situations where multiple teams brainstorm and evaluate ideas for a common purpose, such as a SWOT analysis for multiple stakeholder groups, AI can be used to cluster and analyze ideas across multiple team sessions to highlight areas of strong agreement and point out areas that may have gone unnoticed.

AI-Generated Ideas

Once all team members have generated their ideas anonymously and before the combined list of ideas is displayed, AI could generate additional "anonymous" ideas into the mix for consideration by the team.

Summary Report

At the end of the brainstorming session, AI can be used to summarize the results, including highlighting key ideas, writing paragraphs, and including charts.

AI-ENABLED DIGITAL OFFICE

The following are areas where AI could be used within a Digital Office to augment teamwork.

Collaborative Documents/ Collaborative Team Notebooks

When teams explore new ideas, business needs, and opportunities, they collect, analyze, and synthesize information from multiple sources. Within the Digital Office, they use collaborative documents and team notebooks to capture and organize them. As team members add new content to their team notebooks, AI can be used to summarize and highlight key points of their work to the current point in time, including comments and feedback regarding a specific topic. Similarly, in creating artifacts for initiatives, such as a charter, scope, quality plan, and communication plan, AI can be used to both prepare and review them.

Tutor

AI can be used as a "tutor" using historical information stored in a Digital Office Complex with a prompting interface. A new member might ask to learn about risks from previous initiatives of a similar nature, approaches to resistant stakeholders, or how to improve a scope statement.

"Initiative Plan" Quality Review

By training a neural network using historical data from Digital Offices of "successful" initiatives, AI can be used as a quality reviewer to score the team's "Initiative Plan" on a scale from "Highest Quality – Ready to Go" to "Poorest Quality – Needs Significant Work to Proceed", including recommendations for improvements. It would provide a quality check of the team's planning documents before "execution" as a technique to increase the likelihood of success.

Early Warning Systems

Early warning systems for serious weather events, such as hurricanes, create awareness for authorities so they can act early to save lives. Similarly, leadership teams benefit by knowing early if one of their initiatives will be going "off the rails" to take corrective action to avoid a disaster. Unfortunately, most leadership teams never receive a warning.

By analyzing team communication patterns using historical data of a Digital Office Complex, AI could create a "team moods" classification scheme,

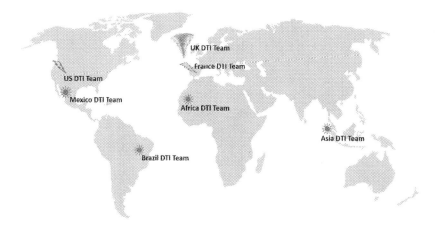

FIGURE 6.2   DTI "Team Mood Map" mock-up.

such as "sunny – performing well", "stormy – slight performance impact", and "hurricane brewing – significant performance impact". The communication of an active team could be compared to this classification scheme to produce a "Team Mood Map" as illustrated in Figure 6.2. It serves as an example of an early warning system. It shows seven teams that are part of a large digital transformation initiative (DTI) plotted on a global map. AI has assessed four of the teams with a "sunny" mood, two with a "stormy" mood, which the teams will need to address, and one where a "hurricane" is brewing. The latter will require urgent attention as it could have a severe impact on the overall initiative.

After the first sip of her morning coffee, Jordan's phone rang. It was her virtual assistant, Nic. One of the things she had Nic tracking was the mood of the teams working on the company's critical "digital transformation initiative". Nic regularly scanned the "Digital Offices" within the "Complex" to calculate a "mood" for each team based on reviewing "Digital Office Objects", such as the Initiative Charter, meeting minutes, Stakeholder Register, communication plan, Risk Register, issue log, change log, schedule, budget, team notebook, and discussions. Based on a variety of internal teams over many years, Acme had modeled a variety of team "moods" to use in detecting potential problems as a part of a proactive approach to avert them, especially those that could become catastrophic.

"Good morning, Jordan. Based on my assessment, there is a significant storm brewing for the UK DTI team". As she was reading the summary report Nic displayed on her screen, she received another call. It was one of the company's board members, Thomas. He was in town and wanted to schedule an update on the "digital transformation initiative". "What timing", Jordan responded. "Would you join me in a call with our UK team? They appear to be facing a sizable challenge. It would be great to have both of us listening to see how we can help".

SUMMARY

The Digital Office Complex organizes both structured and unstructured data for vision delivery for the enterprise. AI, especially the power of neural networks, can be leveraged to augment teamwork. It can be used to assist teams in digital brainstorming and team decision support, preparing an initiative before executing it, and monitoring "in-flight" initiatives to look for early warnings that can impact its performance.

NOTES

1 A. L. Samuel, "Some Studies in Machine Learning Using the Game of Checkers", *IBM Journal of Research and Development*, 3(3), 210–229, 1959.
2 Warren McCullough and Walter Pitts, "A Logical Calculus of the Ideas Immanent in Nervous Activity", *Bulletin of Mathematical Biophysics*, 5, 115–133, 1943.
3 Frank Rosenblatt, "The Perceptron: A Perceiving and Recognizing Automaton", Cornell Aeronautical Laboratory, Inc., Report 85–460–1, January 1957.

Implementation 7

INTRODUCTION

Reengineering a company's vision delivery process is itself a "big idea". It will impact the entire organization. Figure 7.1 shows the flow of changes beginning with the "Digital Office Complex" (1), which is a digital infrastructure architected to enable teams to work together in a unifying manner, replacing the "Islands of Digital Teams" (2) to enable the transformation of "Teaming" (3), including the following:

1. How teams can use digital brainstorming to generate potential goals and evaluate them to prioritize their goals and then track them in their Digital Office.
2. How teams can use their Digital Office work scheduling capabilities to support their coordination needs.
3. How teams can use digital brainstorming and decision support can generate and evaluate ideas from team members in support of divergent/convergent thinking to explore many possibilities and then narrow to a decision.
4. How teams can use team notebooks to develop the team's knowledge by sharing discoveries, findings, and hypotheses and create team "work products".
5. How teams use the approval workflow for internal team support as well as external organizational requirements.

Digital teaming enables reengineering the Vision Delivery process from a fragmented process and disparate data to an integrated process with data stored in a database (4). The Digital Office Complex supports a newly reengineered Vision Delivery process to enable teams to improve (5):

- Clarity of vision, objectives, scope, and results.
- The engagement of stakeholders to participate at any time.
- Prioritization of challenges and work.

  DOI: 10.1201/9781003426127-7

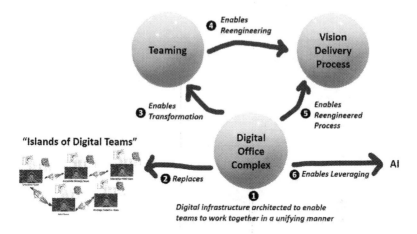

FIGURE 7.1   Digital Office Complex change map.

- Commitment from team members to goals and teamwork plan.
- The visibility of teamwork and teamwork performance by leadership and stakeholders.
- High-quality work products, outcomes, results, and intellectual assets.
- Success for change through the adoption of new technologies and new processes.

All of these are relevant to achieving strategic results. The Digital Office Complex enables leveraging AI (6) to enhance performance with capabilities, such as idea clustering, initiative plan quality review, tutor, idea evaluation results feedback, and early warning system, which contribute to shrinking time to value, or time to bail.

CRITICAL SUCCESS FACTORS

The following are critical success factors common with many high-visibility initiatives.

Championship, Sponsorship, and Participation

Having leadership team members actively and enthusiastically communicate the vision and value of the initiative informs the entire organization of its priority and significance. Sponsoring a cross-functional team to lead the imple-mentation initiative and providing them with the necessary resources is crucial. Ensuring leadership team members' availability to the cross-functional team for their input, feedback, and guidance is critical to the realization of the vision.

Organizational Culture

Beyond the usability of the technology is the compatibility of the organiza-tion's culture, which is its vision and values that direct people in how to behave within the organization. To ensure successful adoption, the organization must value the following:

- Teams. A team is two or more people who have a mutual objective and must coordinate their work to achieve the objective.
- Information sharing. Team members need information from each other to accomplish activities. The contrast is information hoarding.
- Collaboration. Team members must value working together to cre-ate deliverables, outcomes, and results.
- Innovativeness. Bringing forward ideas for new products, services, and improvements.

Clarity of Vision, Scope, Pace, and Results

Through ongoing discussions of the vision, such as describing and responding to example cases, serves to refine it and improve what it means to the organiza-tion. Having a better understanding of the vision enables the cross-functional team to be clear in its scope and the intended results. With ongoing dialogue with the leadership team, the cross-functional team can set a pace that aligns with the desires of the leadership team.

Talent

Identifying the knowledge, experience, skills, and expertise necessary is the foundation of team formation. Team members must be able to "lead" at times with their unique capabilities and "follow" at others. They will need to par-ticipate as a team member in problem-solving and decision-making sessions. They will need to be able to interpret abstract descriptions of the vision and turn them into outcomes.

Insurgents and Incumbents

To reengineer the vision delivery process, it is critically important to include "insurgents". A team of only "incumbents" will most likely just "automate the cow paths" or just "see new things from the perspective of the existing work", while an "insurgent" will embrace "the new and see it as different".

In 1994, while recruiting on behalf of Microsoft at his alma mater, Cornell University, Stephen Sifnosky became trapped in a snowstorm. While waiting, he took the opportunity to observe students creating a software experience using free software from other universities. To him, this was on a direct collision course for Microsoft's grand vision of "Information at Your Fingertips". He captured his thoughts in an email to Bill Gates that led to an offsite meeting with top executives in April 1994 that then led to the famous Gates' email of May 26, 1995 – "The Internet Tidal Wave". In a blog post in 2021, Sifnosky reflected on his experience.

One of the things I concluded was that as I showed different aspects of the internet to Bill (gopher, WWW, ftp, telnet, HTML, etc.) he was quick to map those to existing or envisioned capabilities in Windows or in Information At Your Fingertips. I was struck by this because, well, I did not see that at all. I saw everything on the internet as totally new and different. I saw everything we had as kind of clunky and unrelated, or at least different. As I reflect on this and now have the benefit of the vocabulary of *disruptive technologies*, I can see how I had an insurgent view of the technology whereas Bill had the incumbent view. As the insurgent, I had nothing to lose and everything to gain by embracing the new and seeing it as different. As the incumbent, the natural inclination is to see new things from the perspective of the existing work.[1]

Commitment to Experimentation

Team members will need to incorporate time for experimentation into their plans. Experimentation may be uncomfortable and may create the feeling of slowing things down, but that is where the "learning" occurs. Trying various approaches using the new technology creates alternative methods that may be far superior to existing ways of working. Discoveries may be a better way to accomplish a task, or they may identify a pathway not to take. Team members must try different techniques, not just taking an old process and doing the same with different technology.

REENGINEERING VISION DELIVERY APPROACH

The leadership team starts by creating the vision for the organization. What will a reengineered vision delivery process look like and mean for their company?

To be an organization that successfully brings "big ideas" to fruition in a timelier fashion than our competitors. An organization that encourages ideas from everywhere and anyone. An organization that empowers teams to imagine and act quickly to move its strategies forward. An organization that rewards those who participate in its success.

Replacing our "Islands of Digital Teams" with a "Digital Office Complex" requires a shift in thinking by the leadership team and throughout the company. The perspective changes from an organization that provides its professional workforce with personal productivity tools that trap data in documents scattered haphazardly across the enterprise to one where teams generate knowledge that is captured in the form of digital assets that need to be managed and leveraged. It will persist long after an employee is no longer an employee. With the potential opportunities that arise from AI, it must be organized in a structured manner and not scattered around the organization. We will be an organization that leverages technology, including artificial intelligence, to augment teamwork and accelerate time to value, or time to bail, on our "big ideas".

Figure 7.2 illustrates the issues with the current state of "Islands of Digital Teams" shown on the left side of the diagram, where each team has an associated server to hold their documents in a file folder structure. Team members share and discuss issues in fragmented email threads. To report status, team members extract data from their documents and create a report that they email. There is little to no visibility into any other team's work. With the shift to the Digital Office Complex, depicted on the right side of the diagram, each team stores and retrieves data from their Digital Office, which is part of the Digital Office Complex. Reports and dashboards retrieve and consolidate data directly from Digital Offices. Team members have complete visibility into other team's work. With the data centralized in a single place, teams can directly leverage AI capabilities.

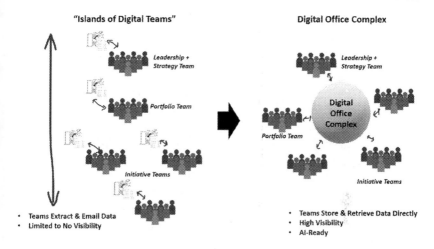

FIGURE 7.2 Change from "Islands of Digital Teams" to the Digital Office Complex.

With the Digital Office Complex, there are new and better ways of working within and between teams. New team practices, or "digital teaming" practices, enable changes to the vision delivery process. The following sections describe the makeup of a cross-functional team to lead the initiative, the use of focus groups and pilots, and a plan to scale.

Cross-Functional Team

The leadership team champions the initiative by sponsoring a cross-functional team that includes members from the leadership team, the strategy team, the project management office (PMO), the human resources (HR) department, the information technology (IT) department, the facilities department, and leadership from the company's business units, departments, and divisions. Their roles and responsibilities include the following:

- Leadership – providing the "vision" for the initiative and championing it throughout the organization. They also set the pace for change.
- Strategy – provides the strategic performance management framework to guide the development of the Digital Office Objects for strategy.
- PMO – provides the portfolio management and initiative management frameworks that will guide the development of the Digital Office Objects for portfolio and initiative management.

- HR – responsible for training both team members and leaders in digital teaming and governing.
- IT – technical implementation and support of the Digital Office Complex, including configuring Digital Offices based on the frameworks from strategy and the PMO.
- Facilities – modify identified "conference rooms" to become "Decision Rooms".
- Business unit, department, and division leaders – champion the implementation effort within their respective unit, including providing requirements to tailor Digital Offices.

The cross-functional team becomes the lead team, the peloton, by using a Digital Office to drive the reengineering initiative itself. They should use Scrumfall, described in Chapter 4, to organize and run it. They will experience digital teaming and governing for themselves. The knowledge they gain from it will give them valuable firsthand knowledge and insights into how they proceed during the implementation process. Each team performance review is an opportunity to share "learnings" that result from directly using the team's Digital Office. What capability did team members use that worked well and could change how the new vision delivery process may work? What challenges did team members have in using the capabilities that will require better instruction for new users? What capabilities are missing and may require development?

As the cross-functional team uses the Digital Office Complex as part of managing the initiative, they should experiment with new ways of working as part of redesigning the processes. Learnings from these experiences will feed improvements to the Digital Office Complex as well as new digital teaming and digital governing practices.

Focus Groups

Focus groups comprised of people from across the organization are organized to test very specific uses of a Digital Office. The purpose is to learn how new users approach the technology and how they learn how to use it. These sessions are one to two hours in length. They involve very specific exercises for participants to complete using a Digital Office and working with other focus group participants. At the end of a session, there is a facilitated discussion with the participants to hear and document their feedback. Following the session, the task is to evaluate how the participants did in learning and using the technology to address a specific application. The output of these sessions is used to improve the configuration of the Digital Offices and provide input for the training strategy.

Pilots

One of the results from the focus groups is a tested and improved Digital Office configuration. Three to five teams working on small- to medium-sized initiatives with durations of three to six months are ideal candidates for pilot teams. It will also include strategy and portfolio teams. These teams will receive training and be provided with support during the pilot with regular feedback sessions to capture their experiences with their Digital Office. Learnings from these pilot teams will contribute to "teaming" practices", "team-to-team" practices", and the reengineered Vision Delivery process with particular attention to the communication between the initiatives, portfolio, and strategy teams. New ideas will emerge that will create faster pathways, more thorough analysis, and better decision-making.

Plan to Scale

Building on these experiences, the cross-functional team will be prepared to create a plan for when, how, and at what pace to scale the Digital Office Complex for the organization. Their major deliverables include the following:

- Digital Office Complex with Digital Offices for teams that are configured with frameworks to support strategic performance management, portfolio management, and initiative management. It should identify standards that are global across the enterprise and those that are regional to a business unit, function, and department. It will also allow for local configuration that is specific to a team.
- Digital teaming and governing practices.
- "Team-to-team" practices for exchanging data between teams, navigating to other teams, and dependency management between multiple team activities.
- Reengineered Vision Delivery process outlining how the flow will work in a Digital Office Complex.
- Process improvement goals that identify the metrics, such as time to value, to evaluate improvements in the vision delivery process.
- Data migration strategy that identifies data from existing initiatives, portfolios, and strategies that need to be migrated into the operational version of the Digital Office Complex.
- Data management strategy that outlines the rules for managing data within a Digital Office Complex.
- Training strategy for digital teaming that outlines the training and materials needed by team members and leaders to become sufficiently proficient in the use of the technology. An example is a training course

for team leaders on how to facilitate a team process that utilizes digital brainstorming and team decision support capabilities.

- Policies that will guide important topics such as security, data ownership, and AI ethics,
- Operations support strategy that defines the organizational support structure needed to operate and provide technical support.
- Benefits plan that will identify the measures that will contribute to improving organizational performance.
- AI experimentation strategy that outlines how the company will experiment with AI capabilities to learn how it can be used to benefit them.

Changing the underlying infrastructure, teaming practices, and the vision delivery process are significant. The organization will need to allow time for the complete organization to adopt and absorb these changes. It will require ongoing support and learning to continue to improve the Digital Office Complex, "digital teaming", and "digital governing".

SUMMARY

With ever-increasing competitive pressures, the need to shrink the time-to-value (or time-to-bail) of a "big idea" has never been greater. Teams produce knowledge and decisions that drive the realization of "big ideas" – new technologies, new products, organizational transformations, healthcare initiatives, or humanitarian development projects. The ability to impact performance depends upon the quality of knowledge being generated by teams in the vision delivery stream and the speed at which it is shared to support decision-making. Unfortunately, the current digital infrastructure for vision delivery teams is woefully inadequate. The "Digital Office Complex" is a collaborative technology environment architected to enable teams to work together in a unifying manner, accompanied by "digital teaming and governing" practices that enable high-performance teaming. Implementing these changes requires a thoughtful and committed leadership team.

NOTE

1 Stephen Sifnosky, https://hardcoresoftware.learningbyshipping.com/p/024-discovering-cornell-is-wired, May 9, 2021.

Printed in the United States
by Baker & Taylor Publisher Services